Deserts and arid lands

REMOTE SENSING OF EARTH RESOURCES AND ENVIRONMENT

Managing editor: V. Klemas

Deserts and arid lands

Edited by

FAROUK EL-BAZ

Itek Optical Systems
Litton Industries, Inc.
Lexington, Mass., U.S.A.

1984 **MARTINUS NIJHOFF PUBLISHERS**
a member of the KLUWER ACADEMIC PUBLISHERS GROUP
THE HAGUE / BOSTON / LANCASTER

Distributors

for the United States and Canada: Kluwer Boston, Inc., 190 Old Derby Street, Hingham, MA 02043, USA
for all other countries: Kluwer Academic Publishers Group, Distribution Center, P.O.Box 322, 3300 AH Dordrecht, The Netherlands

Library of Congress Cataloging in Publication Data

Main entry under title:

Deserts and arid lands.

(Remote sensing of earth resources and environment ; 1)
Includes index.
1. Deserts--Remote sensing. 2. Arid regions--Remote sensing. I. El-Baz, Farouk. II. Series.
GB611.D47 1983 550'.915'4 83-8079
ISBN 90-247-2850-9

ISBN 90-247-2850-9 (this volume)
ISBN 90-247-2851-7 (series)

PRINTED IN THE NETHERLANDS

Preface

Remote sensing is the study of a region from a distance, particularly from an airplane or a spacecraft. It is a tool that can be used in conjunction with other methods of research and investigation. This tool is especially applicable to the study of the deserts and arid lands of the Earth because of their immense size and their inaccessibility to detailed study by conventional means.

In this book examples are given of the utility of aerial photographs and space images in the study of semi-arid, arid, and hyper-arid terrains. Emphasis is placed on the physical features and terrain types using examples from around the world. The authors I have called upon to prepare each chapter are renowned specialists whose contributions have received international recognition.

To the general reader, this book is a review of our knowledge of the relatively dry parts of the Earth, their classification and varied features, their evolution in space and time, and their development potentials. To the specialist, it is a detailed account of the deserts and arid lands, not only in North America, but also their relatively unknown counterparts in North Africa, Australia, China, India, and Arabia.

This book would not have been possible without the dilligent work of my former staff at the Smithsonian Institution's Center for Earth and Planetary Studies. My thanks go to Ellen Lettvin who translated two chapters from the French and assisted me in editing the other chapters. Rose Aiello and Lesley Manent helped me in the preparation of all the illustrations. Special thanks are to Donna Slattery who expertly typed all the versions of the manuscript.

Lexington, Mass. Farouk El-Baz

Contents

1. The desert in the space age

Farouk El-Baz

Abstract

The scarcity of basic information on the arid lands of the Earth has resulted in part from the difficulty and harshness of desert travel. Space photographs and images provide a new tool to study desert features and select areas for detailed field investigation. Color photographs obtained by the astronauts clearly depict variations, which are indicative of the chemical makeup of desert surfaces. Landsat images are useful in the study of temporal changes because of the repeated coverage of the same area at the same scale. Orbital photographs and images show the sand distribution patterns in desert areas and the effect on dune morphology of local topographic variations. Examples are given particularly from the Western Desert of Egypt, the driest part of the North African Sahara. Space age technology also allows the use of automated stations to gather meteorologic data in remote and inaccessible regions. Such advances help us to better understand the deserts of the Earth and better interpret similar terrains on the surface of Mars.

Introduction

Unlike other types of terrain, the driest one-fifth of the land area of the Earth is still shrouded in mystery. Basic information on the desert remains scarce. The main reason for this scarcity or lack of knowledge may be the fact that earth science had its origins in Europe, the only continent without a desert [1]. Pioneers in the understanding of terrain wrote that landscapes are basically sculptured by surface water and ice. Wind-scoured terrain received little or no attention in the classical geologic literature.

Furthermore, 'until recently desert travel has been so difficult that only the hardy traveler would attempt it. Impeding scientific exploration were the vastness of the Sahara, the remoteness of the central Asiatic deserts, and the absence, in many desert areas, of water and vegetation to support men and animals. Since

El-Baz, F. (ed.), Deserts and arid lands. ISBN: 90-247-2850-9.

there appeared little economic justification for such exploration, few studies were made in the past' [2, p. 3].

Few experts agree on one definition of the desert. However, most people agree that the desert is a land area that receives less than 25 cm of precipitation per year. It cannot, therefore, hold much vegetation and remains dry most of the time. This definition includes the polar ice of the frozen tundra, which is characterized by perpetual ice and snow cover and intense cold. Regions of 'cold deserts' include one-sixth of the landmass of the Earth, over 23,000,000 km^2. However, the term desert is more commonly applied to the hot and dry regions of the Earth which are concentrated between 10 and 40 degrees north latitude and 20 and 30 degrees south latitude. These are belts of land fairly close to the Tropic of Cancer and the Tropic of Capricorn, where rainfall is scarce and where the wind is very dry. The 'hot deserts', or hyperarid lands, encompass 31,000,000 km^2, or one-fifth of the land area of the Earth.

The perspective from space

Photographs from space provide a new tool of desert study. The first photographs were taken in 1965 by the astronauts, who did not fully realize the enormous usefulness of the new vantage point. Because of their large areal coverage, these orbital photographs are especially helpful in mapping regional patterns of sand distribution (Fig. 1), in studying large-scale dune morphology, and in determining the direction of sand movement [3–7]. In addition, deserts are particularly suitable for study in orbital photographs because the general lack of cloud cover over arid regions improves image quality. Study of space photographs allows the making of preliminary surveys, which include description of desert landforms and selection of areas for detailed fieldwork.

Space photographs also allow the monitoring of changes to the terrain, which indicate natural improvement or degradation of the environment over a period of years or decades. Photographs taken of the same area show changes to this terrain with time. For example, in the late 1950's the Egyptian Government started 'The Liberation Province', a major reclamation effort just west of the Nile Delta. The project was met with many difficulties during the early years, but as reclamation proceeded northward, the project began to bear fruit. The reason for this became evident after comparing two photographs of the area (Fig. 2), the first taken by the astronauts of Gemini 5 in 1965 and the second by the Apollo-Soyuz in 1975 [8]. The reclaimed area totaled 1,100 km^2 for the ten year period. The photographs showed that the southern part of the project extended into a zone of active sand, which should have been avoided. The northern extension of the project was in more suitable, fertile soil.

Furthermore, photographs taken from space clearly depict subtle variations in

Fig. 1. Landsat view of the Turpan Depression in northwestern China. Sand is confined to a dunefield, 60 km in length, in the eastern part of the depression. Two prevailing northwesternly and northeasternly winds result in two dune forms, transverse chains and pyramid dunes. Fan-shaped lineations along the northern slopes of the depression are caused by running water from melting snow.

the reflectance properties of desert surfaces, which are indicative of changes in mineralogical composition. These photographs have been used in the past to confirm field observations of variations in the red color of desert sands. For example, in the Namib Desert of Southwest Africa, where linear dunes have migrated from west to east along the coast of Namibia, the sands farthest inland are much redder in color and are of greater age [9]. Skylab 4 photographs of the same region show color zones in the dune sand. In these photographs, younger

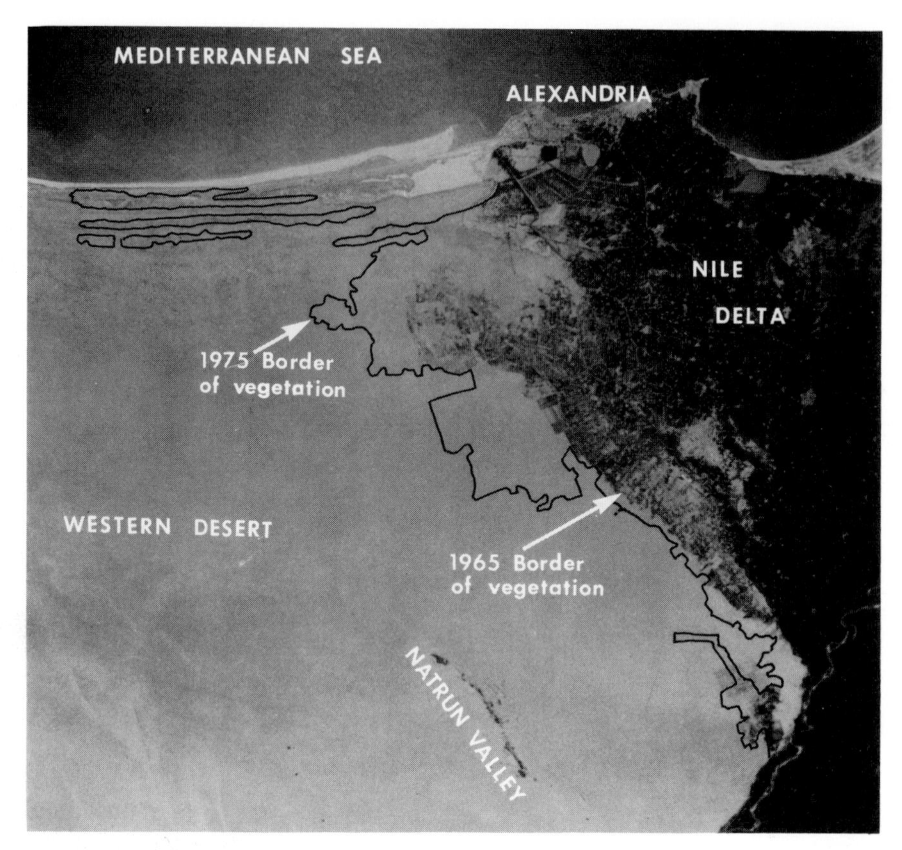

Fig. 2. Illustration of the use of space photographs in monitoring long term variations along the desert edge. Shown in this case is the increase in vegetation by reclamation of land at the western edge of the Nile Delta. Drawn on Gemini photograph S65–45736 are the vegetation boundaries as depicted on Apollo-Soyuz photograph AST-16-1257. The reclaimed area in 10 years is about 1,100 km^2.

sands near the coast appear brighter than the redder zones farther inland [3]. Similarly, Apollo-Soyuz photographs of southern Australia show dune redding as a function of increasing distance from the source [10]. Photographs of the Lake Blanche area in the Sturt Desert and of the Lake Eyre region in the southern Simpson Desert show an increase in red color as the distance from the sand source increases (Fig. 3).

The presence of hematite coatings on individual grains [11] has been given as the cause of the red color in desert sands. However, the origin of such reddening remains controversial. The variation in grain coating is significant because of its effects on the spectral signature of the sands on space images and photographs. Recent studies indicate that the coating on sand grains is composed of kaolinite with powdery hematite, thus linking such reddening to desert varnish [12].

Fig. 3. Apollo-Soyuz photograph of part of the Sturt and Strzelecki Deserts, South Australia. The left edge is occupied by the foothills of the North Flinders Mountains, on the edge of which lies the dry Lake Blanche, which is 25 km wide. The dune sand near the source is light (buff) colored and is dark (red) colored away from the lake and toward the upper right.

Kaolinite in the coating is believed to originate as dust that percolates through the sand and adheres to the surfaces of the grains. Thus, the processes that result in the darkening of rock surfaces with time through the formation of desert varnish also act upon individual sand grains, affecting their spectral reflectance [13].

Detailed studies of sand samples from the Western Desert of Egypt show that the coating on quartz grains in samples from three locations increases in thickness from about 0.5 microns in the north to about 1.5 microns in the middle to between 2 and 5 microns in the south [14]. These results not only confirm the observation of reddening of the sands as the transport distance increases, but also suggest that the quartz grains acquire the coatings during aeolian transport. Therefore, the reddening property can be used to determine the relative ages of color zones in the same sand field. This property can be used to indicate the transport direction of the sand in orbital photographs.

Because of their usefulness, additional color, high-resolution photographs of desert regions are planned on the Space Shuttle missions. This will be done by the use of the Large Format Camera (LFC) to acquire mapping quality, stereo, 10–20 m resolution photographs on color film. The camera derives its name from the size of individual frames, which are 23 $\times$ 46 cm. It has a 305 mm, f/6 lens with a 40 $\times$ 70° field-of-view [15]. The camera is to be placed in the Shuttle's cargo bay: In this mode of deployment the LFC operates in Earth orbit, returns with the Shuttle, and provides the photographs rapidly and cost effectively. From the Shuttle altitudes of 200–400 km, the coverage of this camera will allow the study of up to 87,000 km^2 in one photograph. Such photographs are ideal for studies relating to the regional distribution patterns of desert sands and the orientation and direction of the motion of dunes.

Results of aeolian action

Running surface water no longer plays a significant role in the transportation and deposition of sediments in the hyperarid deserts. Deserts are usually dominated by aeolian activity, in which wind is the major agent of erosion, transportation and sedimentation. In this environment, particulate material originates from the disintegration of rock by both mechanical and chemical weathering. The disintegration exposes the particles to the agents of erosion, which in turn cause more loosening of particles from the rock.

Wind deposits are discretely zoned in response to the capacity of the wind to sort out and segregate particles by grain size. The finest particles (clay and silt particles, up to 0.05 mm in size) are winnowed out wherever they are exposed and whirled away into the atmosphere as dust. The dust particles settle out of suspension beyond the zones of high wind energy [16]. Saharan dust storms, for example, carry such fine dust particles concentrated at altitudes of 1.5 km and 3.7 km in the atmosphere. It was estimated that 25 to 37 million tons of dust are transported through 60° W longitude each year. This is equivalent to the present rate of pelagic sedimentation for the entire north equatorial region of the Atlantic Ocean [17]. Similarly, in temperate regions the wind picks up particles of glacial silt and clay, and deposits them as loess. A buff colored deposit of loess up to 100 meters thick, wind-transported from the Mongolian Plateau covers most of northern China. Similarly, loess from the Great Lakes region and from Canada covers much of the central United States.

Fine to medium sized sand grains bounce or saltate readily in strong desert winds. By the saltation process, the wind segregates particles of 0.05 to 0.5 mm size from the clays, silts and gravels, and shepherds them into dunes. The movement of sand by saltation depends on the nature of the surface. The sand particles jump to higher levels if they saltate on gravel than they do on a sandy surface.

Larger particles (0.5 to 2 mm in diameter). too large to be lifted off the surface by the bombardment of saltating grains, may gradually and erratically move or roll along the surface. With high winds the whole sandy cover appears to be creeping slowly on the surface along the wind direction.

Wind-induced separation of particles into varying sizes results in the formation of vast flat plains that are veneered with a surface of granules or pebbles, usually well-sorted and one grain thick (Fig. 4). The lag deposit forms an armor, which is usually in equilibrium with strong winds. The process of surface creep, by which the wind distributes the coarse materials seems imperceptible except under rare high-velocity wind conditions. Removal of the coarse lag produces lasting scars that result from deflation of the lag-protected silts and sands.

Sand-sized grains accumulate in the form of sand sheets, usually with rippled surfaces or as free and obstacle-attached dunes. Free dunes, which are by far the most abundant, can be isolated simple forms, coalesced compound groups, or complex accumulations. The basic geometric shapes of dunes are straight (linear, transverse, and also linear dunes), crescentic (barchan and parabolic dunes), and domical (including dome, pyramid and star dunes). All types of dunes may be active, semi-fixed or fixed by vegetation [1]. Dune shapes vary considerably in different areas and often within the same locality (Fig. 5).

A group of dunes may form a bouquet or bundle, larger numbers a dune field, and huge accumulations of sand dunes are known as sand seas [18]. Detailed knowledge of the texture and morphology of these sand seas was made possible through the interpretation of photographs and images taken from space. For example, data from Meteosat allow the recognition of general sand movement patterns in all of the Sahara. Similarly, Landsat images and photographs taken on manned space missions allow the mapping of dune accumulations such as those of the Great Sand Sea in the Western Desert of Egypt and of

Fig. 4. A desert pavement with a protective layer that is one grain thick in the southern part of the Western Desert of Egypt. The largest grain in the center of the photograph is 2.5 cm long. The surface grains protect the sand beneath from wind erosion.

the Taklimakan Desert in northwestern China.

Wind erosion creates long corridors in the hardest rocks. Such corrasion features, shaped solely by the wind can be 100 km in length [5]. Yardangs, the smaller wind-sculpted features that resemble inverted boat hulls, occur in numerous wind eroded deserts. The role of vorticity in developing small scale lineations by wind erosion of rocks is also well established [19, 20].

Sand distribution and motion

As stated above, space photographs are very useful in establishing the distribution of dunes. With repeated coverage of the same area, the photographs may be utilized in recognizing the direction and perhaps the rate of motion of dunes. For example, Apollo-Soyuz photographs and Landsat images were used in establishing the distribution pattern of sand dunes in the Western Desert of

Egypt [21]. Furthermore, Landsat images were used to map dune lines in the Kharga Depression [22].

The Kharga Depression is an elongate hollow oriented in a roughly north-south direction (Fig. 6). It ranges from about 20 to 100 km in width and is bordered on the north and east by a steep disjointed scarp about 200 m high. On the west it opens onto the Dakhla Depression. The southern boundary is not well-defined, as the depression floor blends gradually into the terrain to the south. A few isolated hills dot the otherwise flat floor. Structurally, depression strata represent a subtle fold oriented nearly north-south which is delineated by beds with very gentle dips to the west-southwest and east-northeast. Running down the middle of the depression is a series of steeply dipping faults [32] along which much of the region's scant vegetation is centered (Fig. 6).

Dune deposits at Kharga were mapped at 1:250,000 scale using a computer-enhanced Landsat color composite as a base (Fig. 7). The dunes appear to be concentrated in the eastern and central parts of the depression and on the plateau to the north. Sand deposits on the depression floor form a pattern of discrete, roughly north-south trending sand streaks which are composed largely of barchan dunes. Sand from the large Ghard Abu Muharik dune on the north plateau pours into the depression and is channeled into separate bands by wadis along the scarp [23].

Because the wind within the depression comes consistently from the north, the Kharga dunes proceed southward, maintaining the form of the individual streaks. The rate of movement of individual barchan dunes over a wide range of sizes in the Kharga Depression was measured in the field [24]. The rate varies between 20 and 100 m per year with an average rate of 50 m per year. Dune motion is strongly related to size and shape of the dune; larger dunes travel slower and smaller ones faster.

With the knowledge of the average rate of motion of the dunes and the unidirectional nature of the wind in the depression, it is possible to forecast dune positions in the future and to determine the places most likely to remain dune-free for extended periods of time. Several such areas were identified on Landsat images [22] and five of them in the field.

As shown in Figure 7 the five sites are located in interdune areas, which are likely to remain free of sand. Site 1, extends about 40 km from the northern encarpment of the depression. Although its southern part exhibits a carbonate-rich soil that is covered by chips of desert varnished sandstone, its northern third is composed of fertile clay-rich soil. The interdune corridor appears to be part of a playa deposit that is nearly 10 km in width. The thick, sparsely vegetated playa soil is, in places, covered with a thin veneer of sand. The presence of the sand increases the porosity, resulting in a soil which, if properly irrigated, would be suitable for agricultural projects such as the introduction of crops for grazing. Field observations indicate that near the north scarp the potentially productive

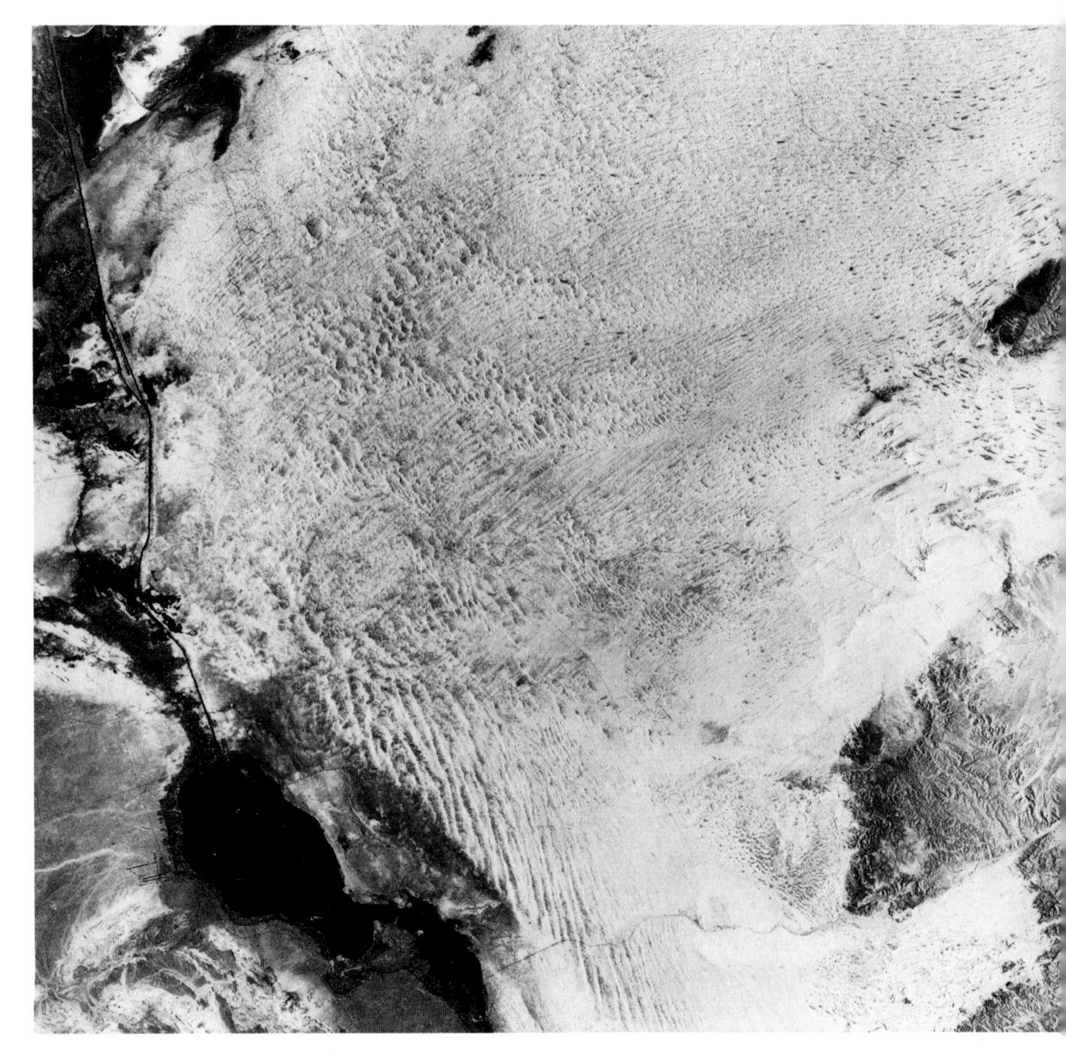

Fig. 5. Part of the sand field in northwestern Sinai Peninsula east of the Suez Canal and the Bitter Lakes; the Great Bitter Lake at lower left is 13 km wide. The pattern of short, partly sinuous dunes and blowouts in the north changes to linear dunes, 10–15 km long, to the south. The eastward wind has channeled the sand into a closed graben in the lower right corner of the Landsat image.

zone is as much as 6 km wide. The presence of meter-size mudcracks indicates the occurrence of occasional rainfall. A nearby spring, more than a kilometer from the northern scarp, suggests the possibility of the existence of retrievable ground-water in the region.

On the other hand, sites 2, 3 and 4, which are closer to the center of agriculture at Kharga, were found to be much less productive. The soil in these three sites is also fertile, but the terrain has been drastically eroded by the wind. In most places

Fig. 6. Computer-enhanced Landsat image of the Kharga Oasis region. The scarp that encloses the depression on the east side is about 200 m high. Light-colored streaks within the depression are composed of barchan dune chains. Irrigated agriculture in village oases appears as blacks patches in a north-south orientation at center. The dashed line is the groundtrack of orbit 15 of the Space Shuttle Columbia. Areas A and B represent sites from which samples were collected by the author at the time of Columbia's 2nd flight (12–13 November, 1981) for later correlations with data from remote sensing instruments onboard the spacecraft.

the terrain is undulating with relatively thick sand deposits in the low areas. Where there is a measurable thickness of fertile playa deposits, as in site 4, the wind carved yardangs three to four meters high and tens of meters long. Only scattered, small patches of flat terrain may be usable in these sites.

The most southerly of the five sites in Figure 7, site 5, by virtue of its size, location, and soil, exhibits excellent potential for the expansion of villages and agricultural activities. There does appear to be a distant threat of dune encroachment but, given the average rate of movement of 50 m south per year, the northernmost reaches of this area should not be affected by dunes that are large enough to be visible on Landsat images for more than a century.

This example shows that the perspective gained from Landsat images of the Kharga Depression allows the relatively rapid identification of areas of dune accumulation and sand-free interdune zones. Furthermore, the remarkably consistent nature of the Kharga winds creates a 'natural laboratory' for observing the interactions of wind-blown sands with barriers of different shapes and properties. The information gained from study of the images, with field checks, is important to the planning of agricultural expansion in the region and to the selection of new village sites. The contribution from Landsat is timely in view of the fact that the Government of Egypt is planning further development of this region as part of the New Valley Project during the coming decades. Thus, the information was recently shared with officials of this region for consideration in their development plans.

Composition of desert soils

The spectral signatures of desert surfaces are easily distinguishable in remote sensing data. Landsat data, for example, clearly show differences in the reflectance properties of limestone rock, clayey deposits, and dune accumulations of quartz sand. The Kharga Oasis region of the Western Desert of Egypt was mapped in this way using Landsat images couplied with field checks (Fig. 7). However, as stated above, a thin mantle of one-grain thick deposits may mask the nature of the soil below (Fig. 4). Additionally, clay-rich desert varnish on limestone or sandstone may give the spectral signature of clay deposits as in the case of the southern part of site 1 in Figure 7. Such distinctions must be made based on field observations.

An experiment was recently conducted in conjunction with the second flight of the Space Shuttle Columbia. Samples were collected from exposed rock rubble, soil and sand in the Kharga Oasis region along the track of the Shuttle. Mylar markers were placed on the desert surface to identify the sampling sites in remotely-sensed data for correlation (Fig. 8). Spectral characteristics of these vegetation-free surfaces were correlated with data from the Shuttle Multispectral Infrared Radiometer (SMIRR) and multispectral data from Landsat (Fig. 6).

The SMIRR consisted of a 19 cm diameter telescope followed by a spinning filter wheel and two thermoelectrically cooled detectors. The instantaneous field-of-view was a 100 m diameter spot directly beneath the Shuttle spacecraft.

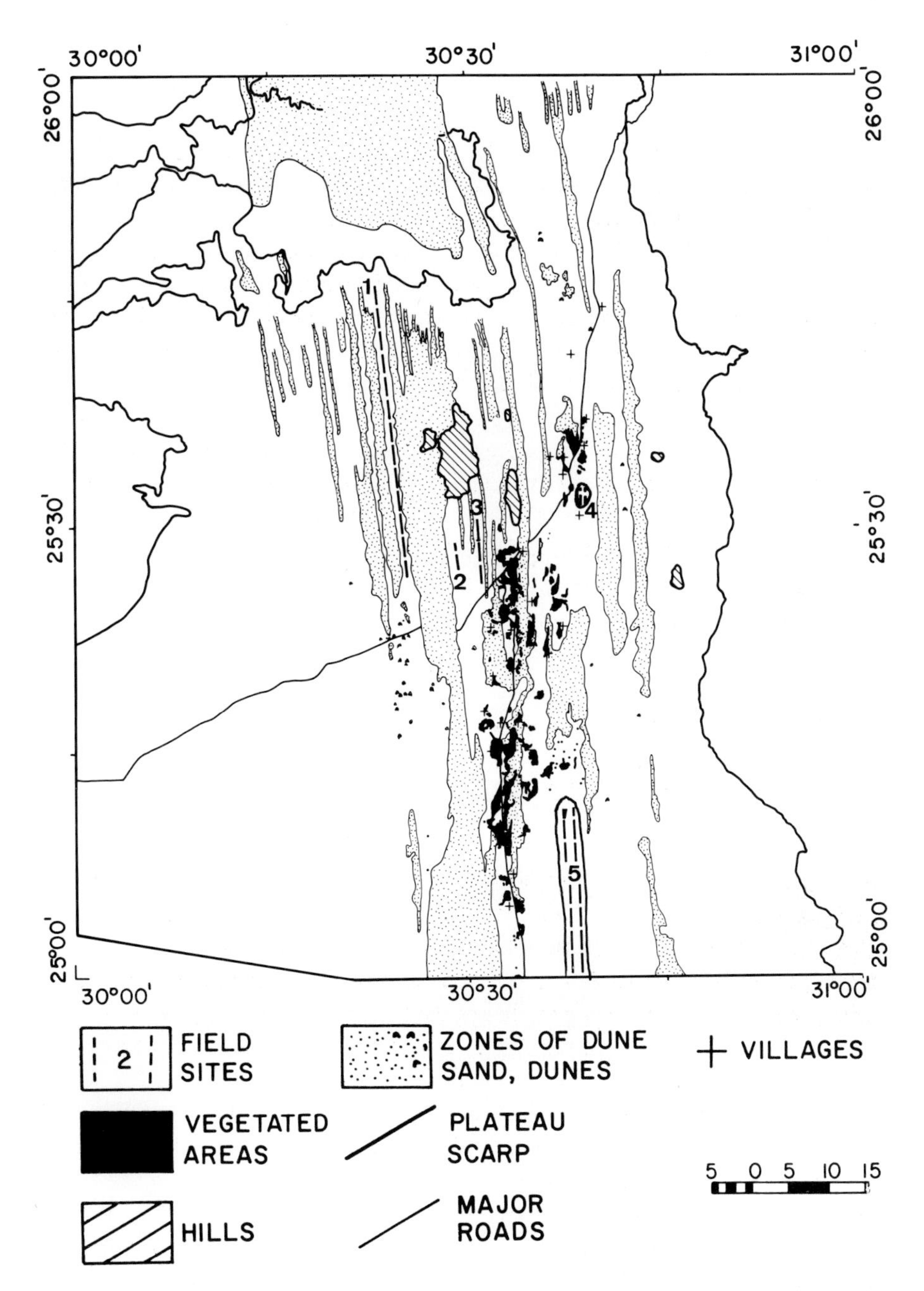

Fig. 7. Map of the distribution of sand dunes in the Kharga region based on the interpretation of Landsat data [22]. The vegetated areas occur along a north-south trending fault. All agriculture is based on underground water wells. The interdune areas expose soils of varying degrees of fertility. The numbered sites have been field checked. The northern part of site 1 and most of site 5 represent fertile soils that are naturally protected from sand encroachment at least for a century.

Fig. 8. A strip of highly reflective 'Mylar' is spread on the desert surface in the Kharga Depression at Site A in Figure 6. The Mylar sheet, 1.5 m wide and 26 m long, was placed to mark the site in a Landsat image obtained shortly after the placement. Because of strong winds, the sheet was anchored to the surface by rocks.

Fig. 9. Ground view of the central part of site 1 in Figure 7. Gebel Tarif breaks the otherwise flat horizon, which is marked by the eastern escarpment of the Kharga Depression. Fertile soils appear as dark areas between sand-covered zones. Sand shadows in the lee of the sparse desert bushes betray the wind direction. In the foreground are ripples on a thin sand sheet. The mixing of soil types within a small area and the thin veneer of sand in places show how misleading it could be to judge the soil composition of interdune corridors in Landsat images. Field checking in such cases is mandatory to the correct interpretation of the terrain.

The position of the spot was determined by two bore-sighted 16 mm framing cameras with glass platens containing fiducial marks. Timing information accurate to 0.01 seconds was recorded on the film edge [25].

The SMIRR detectors obtained radiometric measurements of the Earth's surface in ten spectral channels between 0.5 and 2.35 μm. The selection of the range was based on results of field and laboratory spectral measurements. These measurements showed that the region from 2–2.5 μm in particular contained diagnostic spectral absorption features for layered silicates and carbonates [25].

In conjunction with the Columbia's mission, samples were collected on 12–13 November 1981 along the groundtrack of orbit 16 in the Kharga region. Two sites were investigated in detail (Fig. 7) and three samples were collected from each. In the field it was noted that the surface cover was composed of a mixture of soils and sands (Fig. 9). Specific areas had to be chosen for the sampling sites, with several hundred meters of the same apparent composition.

Spectra of the collected samples were measured in the laboratory to correlate with orbital data from SMIRR. This allowed the recognition of the major mineral components in the soils at the investigated sites. Notable among the identified components are carbonates, clays and quartz sands. Soils with mixtures of these components were also detected. The success of this investigation showed that remote sensing techniques may be developed in the future to identify rock and soil compositions from Earth orbit.

Topographic effects on wind direction

The regional view from space images allows the examination of the various effects of natural barriers on wind direction. A detailed investigation of such a relationship was made in the Western Desert of Egypt in the course of mapping the distribution pattern of dunes [21]. In this desert, the general dune orientations change from N-NW in the northern part to N-NE in the southwest [23]. Many dunes are intimately associated with the scarps that bound 6 depressions [1]. This relationship is believed to result from the interaction between sand-carrying winds and scarps, and other topographic features [21].

In the Kharga Oasis (Fig. 6) it is clear that several types of natural barriers affect the distribution of sand streaks. Vegetation, which effects the wind pattern and velocity, interrupts the regular dune streak configuration. A few isolated hills also act as obstacles to the wind-transported sand. The most notable of these are Gebel Tarif which presents a broad face to the oncoming sand, and the narrow Gebel Ter which is elongate in a north-south direction. Streamlined hills generally divert encroaching sands around either side whereas broad obstacles do not deflect the sand, but merely interrupt its progress briefly until it works its way up and over the barrier and continues on its previous course [26]. This is clearly visible in the upper left part of Figure 6.

Not only do prominences and hills affect the sand distribution pattern, but also topographically low areas influence the dunes. The wadis along the northern Kharga scarp channel both wind and sand [23], and initiate the pattern of discrete streaks of dunes with alternating interdune corridors (Fig. 6).

In the Western Desert of Egypt spindle-shaped, dark streaks form in the lee of topographic barriers. This is particularly evident in the Uweinat Mountain region in the southwest corner of the desert at the intersection of the borders of Egypt, Libya and Sudan (Fig. 10). The Uweinat Mountain itself is nearly 45 km in width. The dark-colored streak in its lee is 72 km in width and 130 km in length [27]. This streak is free of sand due to the topographic shadowing by the Uweinat Mountain.

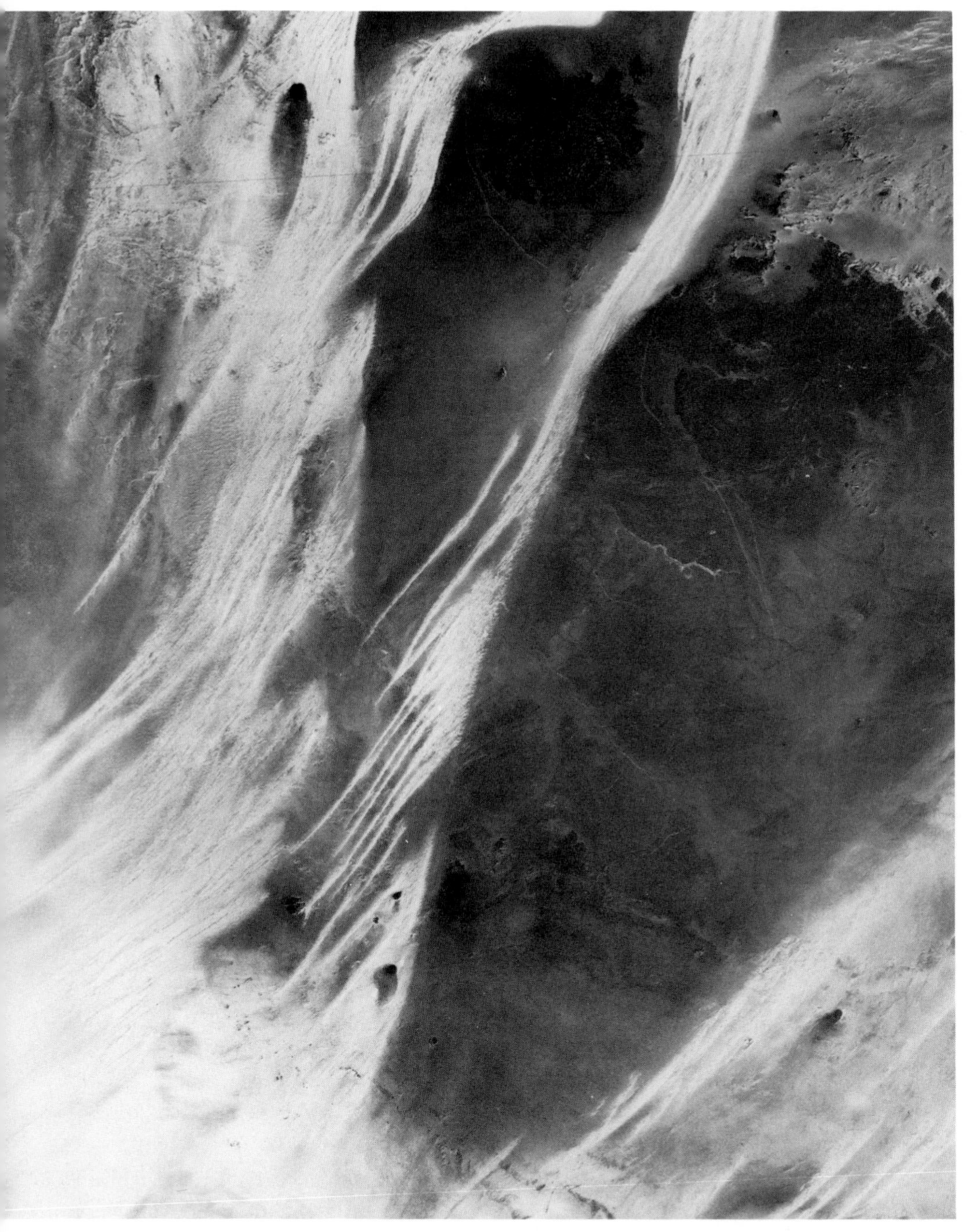

Fig. 10. Landsat image of the Uweinat Mountain region at the borders between Egypt, Libya and Sudan. The width of the image is about 180 km. The dark, spindle-shaped areas are sand-free zones in the lee of topographic highs. The largest of these zones is the shadow of the Uweinat Mountain. The sandy corridors between the mountains are occupied by seif dunes. The orientation of these dunes is visibly affected by the local topographic setting.

The strongest effect on wind direction, however, may be imposed by high escarpments. This is the case in the Farafra Depression in the central part of the Western Desert. This depression is bounded by three scarps creating a roughly triangular shape. It is enclosed on the northern side by a low scarp and on its eastern and western sides by higher scarps. The distance between the east and west cliffs at the latitude of the Farafra Oasis, N27°, is about 90 km. The El-Quss Abu-Said plateau west of the oasis is oriented N45° E with its long axis about 63 km in length, and its width about 29 km (Fig. 11).

The Great Sand Sea west of the Farafra Depression is composed of whaleback dunes with small linear dunes superimposed on top. The average length of the dunes is 70 km, and the average width is 2.5 km. The dunes vary in orientation, between N-S to N26° W, with an average orientation of N13° W [28].

East of the Great Sand Sea in the Farafra region is a smaller dunefield between El-Quss Abu-Said plateau and the northern escarpment (A in Fig. 11). Within this dunefield, the size and orientation of dunes are different from those of the Great Sand Sea. Here, the dunes are smaller, with an average length of 7 km and an average width of 0.2 km. They converge and taper to the northeast and are more widely spaced to the southwest. The orientation of the northern scarp is N58° E and of the Abu-Said plateau is N45° E. The resultant direction of these two scarps is N52° E, which is close to the orientation of the dunes, being N50° E [28].

West of this area is another smaller dunefield that is oriented N6° E (B in Fig. 11). It is surrounded by two scarps, one oriented N50° E and the other N38° W. The resultant direction of these two scarps is N6° E, which is parallel to the orientation of the dunefield. Comparison of the size and orientation of the Great Sand Sea dunes and those in the Farafra depression dunefields suggests that the topography is affecting the direction and strength of the wind in the formation of dunes within the depression.

The available wind data were studied to correlate them with the dunes in the Farafra region. These data were derived from surface wind N-summaries recorded in the Farafra oasis between 1958 and 1966, and were compiled and prepared by the Environmental Technical Applications Center of the U.S. Air Force. The wind speed was recorded in knots to the nearest 10° of direction, at 3–6 h intervals. During this nine year period, a total of 11,844 observations were recorded, of which 2,683 were calm winds (less than 1 knot). The rest were calculated into percentages of the total amount of wind and then organized by month. Each month was broken up into sixteen directional sources which were grouped into five categories of velocity.

To determine sand-moving winds and drift potentials, at a given height, the threshold wind velocity must be known. The threshold wind velocity at the height is the minimum velocity at which the wind will set the grains in motion. This value, as calculated from Bagnold's equation [23], is 11.6 knots. This falls

Fig. 11. Map of the distribution pattern of sand dunes in the Farafra Depression in the central part of the Western Desert of Egypt. The regional pattern of dunes in the region is exemplified by the orientation of dunes of the Great Sand Sea at the left edge. This pattern is quite different from those of dunes in areas A and B within the depression, which are controlled by local topography.

into the 7–16 knot velocity category on the N-Summaries. In order to find the amount of winds above the threshold wind velocity at ten meters high, a percentage of the amount of wind in each direction in the velocity category 7–16 knots and 2/3 of the 7–16 knot velocity category was calculated to plot the frequencies of sand-moving winds.

In Figure 13, the annual summary of winds, basically from the north, is given on the right side. This summary can be misleading. When sand-moving winds are plotted by season and month (Fig. 12), changes in wind direction are apparent

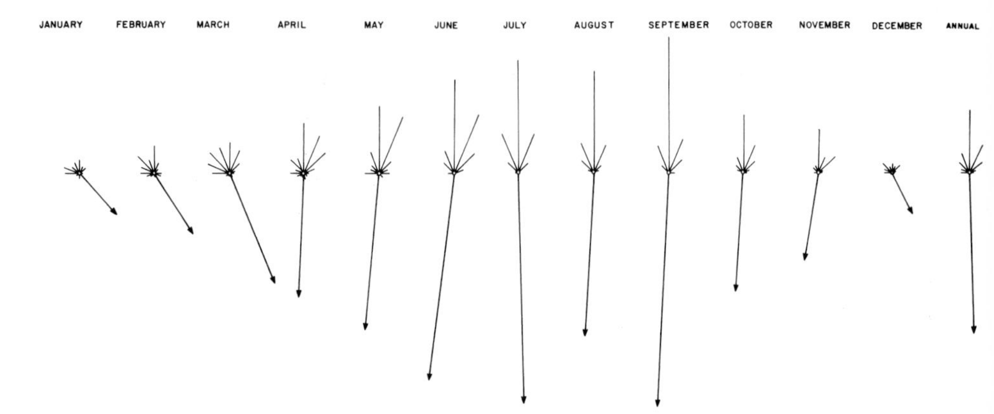

Fig. 12. Summary of winds as recorded at the Farafra Oasis meteorological station between 1958 and 1966. The direction of resultant winds are shown by arrows for the twelve month of the year. The annual summary, of basically northerly winds, is shown at right.

throughout the year, and agree with the dune patterns in the Great Sand Sea. However, there is not a single month with wind blowing in the direction of the dunes within the Farafra Depression. Therefore, the scarps must affect the orientation of the dunes by deflecting the wind. The resultant directions of the scarps that bound the two smaller dunefield are nearly the same as the orientations of the enclosed dunes [28].

These examples show that although dune orientations give clues to wind direction, in some instances these orientations are affected by local topography. Therefore, distinctions must be made between the regional wind patterns and the local topographic setting. This is very significant in the selection of sites for meteorological stations in desert regions.

The example also illustrates the need to obtain wind data in numerous localities in the open desert. Space-age technology has paved the way for the utilization of automated stations that can gather meteorological data in remote desert regions. Data collected by such means can be transmitted to orbiting satellites, which then retrasmit the data to ground receiving stations for processing, distribution and analysis. There are three basic elements to such a scheme: (1) a data collection platform connected to the sensor recorders; (2) a radio transponder with receiving and transmitting capabilities on board a satellite; and (3) a data receiving station for retrieval, processing, and dissemination of the collected data.

Formation of sand seas

It is now well established that desert regions have evolved under alternating cycles of dry and wet climates [1, 7, 20, 27]. However, contrary to widespread belief, particulate materials in terrestrial deserts are formed basically during wet climatic periods. During dry conditions particulate materials are transported or redistributed by the action of wind. Negligible amounts of particulate materials are formed by wind erosion as compared to those produced by water erosion. Such distinct roles of fluvial and aeolian processes in the generation and redistribution of particulate materials in terrestrial deserts is important to our understanding of the formation of sand seas.

Cursory observation of the distribution of sand seas in the world indicates that sand accumulations occur in basins or low topographic areas. Examples include the sand fields of the Simpson in Australia, the Taklimakan, Gurbantunggut and Turpan in China (Fig. 1), the Rajasthan and Thar in India and Pakistan, the Dasht-e-Lut in Iran, the Great Ergs and Murzuq in the Sahara, the Vallecito and Marayes in Argentina, and the Algodones in California.

Most of these basins are nearly circular or elongate in shape. Some of them are surrounded by high mountains, and most of these show relict drainage leading to the sand fields. This suggests that the sands that fill the basins were produced by fluvial action and deposited in the low areas. Such sand deposits may be 150 meters thick, such as in the Rajasthan of northwest India. In the latter case aeolian activity appears to have affected only the uppermost horizons where parabolic dunes occur over a thick layer of sand. These dunes are partly stabilized by natural vegetation.

The Great Sand Sea was so named by the German explorer Gerhard Rolfs who explored parts of it in 1874 and described it as sand dunes with sand in between, where at distances of 2 to 4 km 'one sand ridge followed the other, and each was over 100 m high' [29, p. 172–173].

The segment of the Great Sand Sea in the Western Desert of Egypt is 72,000 km^2 [21]. However, several additional arms to the sand sea in the same desert may also be considered part of it. These include the sand fields of Sitra (2,100 km^2), El-Quss Abu-Said (800 km^2) and Uweinat (31,500 km^2), giving a total of 106,400 km^2.

The Great Sand Sea begins in the northern part of the Western Desert just south of El-Diffa plateau. It continues southward for 600 km before the dunes are deflected by the Gilf Kebir plateau and associated highs. In the northern part there are few sand-free areas. Numerous 2–10 km long crested linear and sinuous dunes arise from a blanket of sand. The distance between the crests of these dunes is usually less than one kilometer.

In the southern part of the Great Sand Sea, large and gently sloping whaleback dunes abound (Fig. 13). These are usually over 20–70 km long and 2–5 km wide

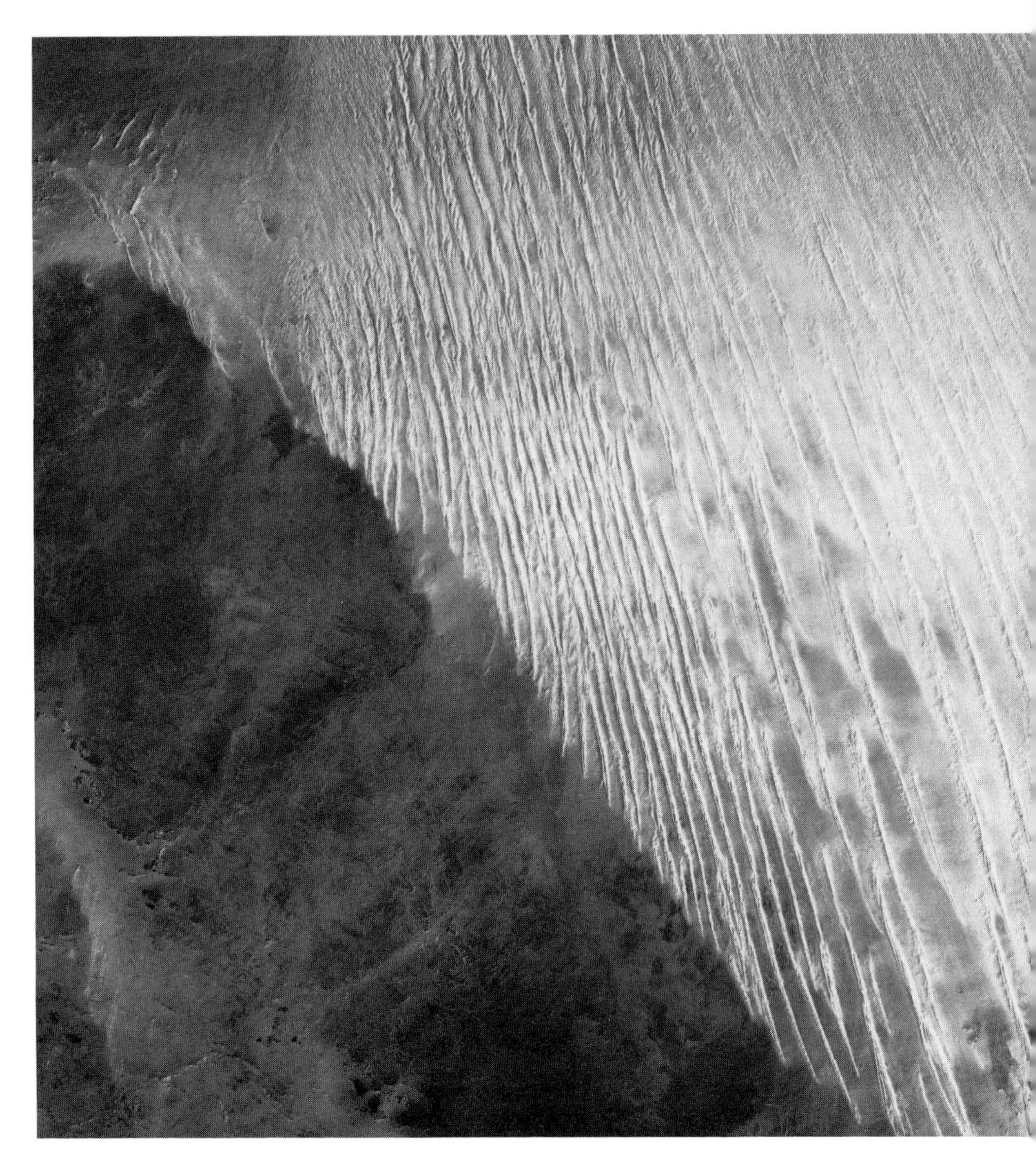

Fig. 13. Landsat image of part of the western border of the Great Sand Sea. The interdune corridors increase in width closer to the southern edges of the sand sea.

[18]. They are often separated by essentially sand free corridors several kilometers in width. These dunes are usually overlain by crested linear dunes or seifs [29], which move along the spines of the relatively static whalebacks (Fig. 14). This led Bagnold to the assumption that the whaleback dunes became so large

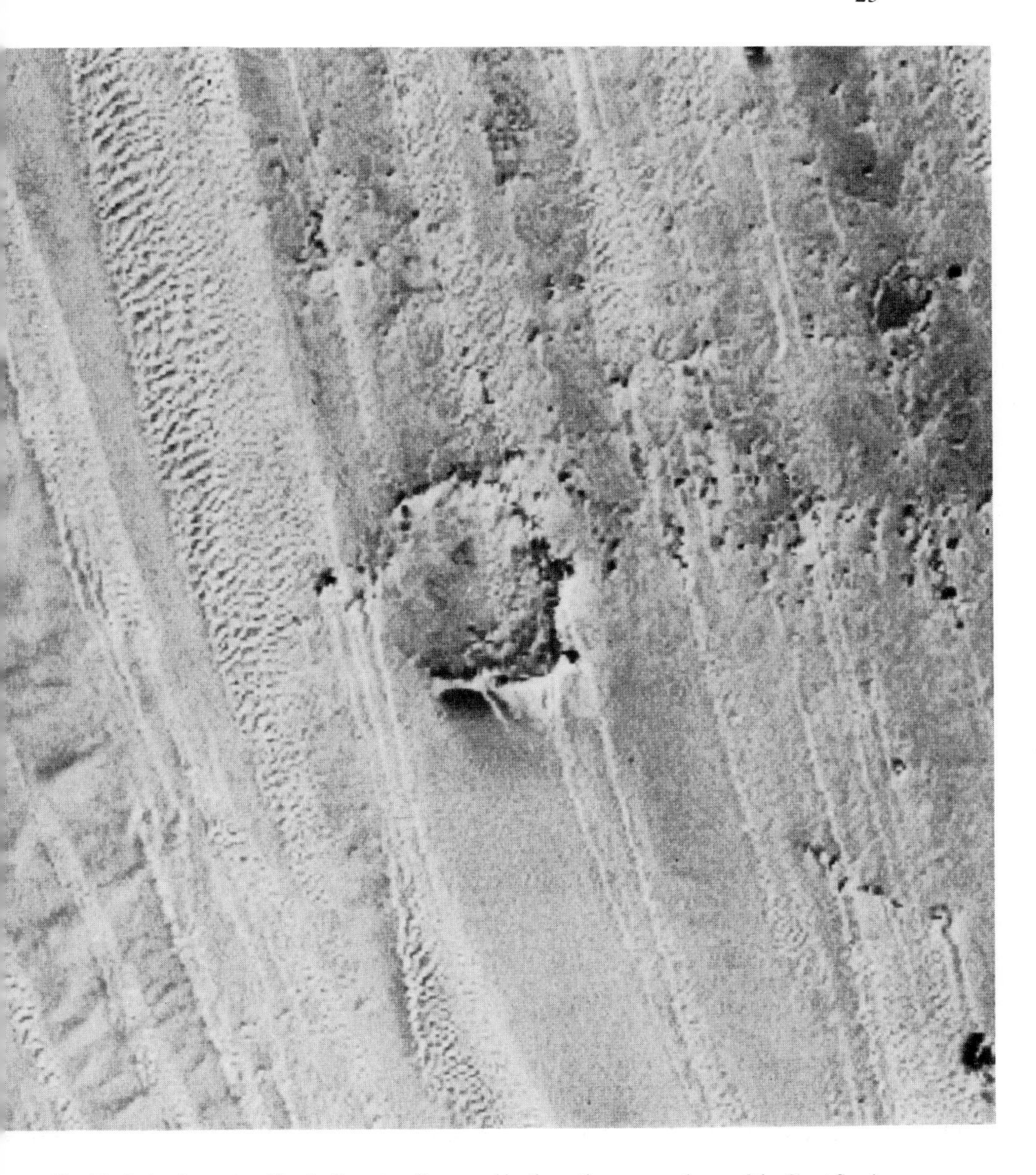

Fig. 14. A circular crater, 4 km in diameter, discovered by the author among dunes of the Great Sand Sea [18]. The crater displays a complex structure with a flat floor, a terraced wall, a crenulated rim, and the subdued remains of an inner structure approximately 1.5 km in diameter. In its interaction with the wind-blown sands, the crater shows similarities to wind-modified craters on Mars. Notable in this regard are the manner in which dunes are deflected by the crater rim, and the presence of a sand-free dark splotch in the lee.

that they no longer could move, and thus, were rendered static by their own bulkiness [23].

However, the situation is better explained if we assume that the whaleback dunes are static remnants of nearly flat (fluvial) deposits of sand that were eroded by the wind. In this model for the formation of whalebacks, sand, over 100 m in thickness, is deposited during fluvial periods in a low basin. During dry periods the sand is moved by the wind which creates the crested dunes of the northern part of the Great Sand Sea. In the southern part, at the edge of the basin where the sand deposit is thinner, the wind erodes the sand more easily and creates the nearly parallel corridors leaving the whaleback ridges. These remain static except for the loss of sand, which forms the sharp crested dunes.

The model clearly explains some of the common features of sand seas, which include: (1) concentration of the sands in basins; (2) the static nature of whaleback dunes; (3) the increase of the interdune distances near the edges of the sand seas. Thus, in dealing with large dune accumulations, distinctions must be made between sharp-crested dunes that result from the transportation of sand by the wind and gently sloping ridges that result from the erosion of sand on either side.

Analogies with Martian features

When the Mariner 9 spacecraft started transmitting images of the planet Mars in 1971, little was revealed by these images. The Martian surface was veiled by an intense global dust storm that persisted nearly two months. After the dust settled, the Mariner 9 images began to reveal a desert-like surface [4]. The images showed that aeolian action was not restricted to raising surface dust. Wind-blown features were abundant, indicating that aeolian processes have played a major role in shaping the surface features of Mars. Mariner, and later the Viking spacecraft obtained evidence of a wind-eroded terrain, a mantling of wind-blown deposits, and wind-related features surrounding Martian craters [1, 18, 20, 27].

In this section analogies are drawn between some of the features of terrestrial deserts and the Martain surface. Examples are given in the Western Desert of Egypt, because it displays the largest number of such analogies. The reason for this is believed to be due to the present day hyper-aridity; it is part of the driest large expanse of desert on Earth [30].

In the hyperarid Western Desert, there are remnants of pluvial-interpluvial cycles including lacustrine deposits, inverted wadis, and terrace deposits along escarpments. Most prominent among the pluvial features are the dry channels in the Gilf Kebir plateau, which is capped by wind-resistant silicified sandstone. Incised into the 300 m high cliffs of the Gilf are 10 to 30 km long channels that end abruptly as box canyons. The lack of significant catchments on the plateau surface suggests that the channels were formed by ground water sapping at the base of cliffs. Another possibility is that former catchment areas were present in softer beds that have been deflated by aeolian action. In both cases, analogies could be drawn to cliff-related channels on Mars.

It is significant that in the open parts of the Western Desert, linear dunes predominate singly, as dune bundles, or in the vast Great Sand Sea. However, crescent dunes occur basically in depressions, particularly along escarpments. Crescentic dune accumulations are analogous to the larger dune mass in the north polar region of Mars, which may be similarly confined, in a low area bounded by a plateau.

A circular crater, 4 km in diameter, among the dunes of the Great Sand Sea displays dark colored material in the southern part of the crater interior. More importantly there is a distinct dark patch in the lee of the crater (Fig. 14). This El-Baz Crater [18] appears to have modified the patterns of material transport by the wind in exactly the same way as craters do in the Cerberus region of Mars.

Dark streaks in the wind shadow of mountains and hills in the Western Desert form streamlined patterns that are similar to those in the lee of craters on Mars (Fig. 15). The spindle-shaped streaks in the Egyptian desert change boundaries in response to changes in the depositional pattern of sand on either side; alterations of the deposition of sand occur due to changes in wind direction.

Unlike shadow streaks which form due to the lack of deposition in the lee of large topographic prominences, knob streaks develop from the deposition of aeolian material downwind of gaps between knobs. In the Sahara as on Mars, these streaks are usually lighter in color than the surrounding surfaces (Fig. 15). Study of the shape parameter in both cases shows a higher degree of streamlining of the terrestrial streaks, indicating a more efficient aeolian regime.

Much like the equatorial region of Mars, the Western Desert of Egypt displays large fields of parallel corrasion features and numerous yardangs [30]. In addition, remnants of scarps in the Western Desert are usually in the shape of inselbergs or knobs. Similarly, knobby terrain on Mars occurs near boundaries between plains and plateau units. The length to width ratios for measured knobs in Farafra (Egypt) and in Cerberus (Mars) are 0.63 and 0.65 respectively, even though the Martian knobs are 100-times larger than those in the Western Desert.

Blocks of varying sizes occasionally litter the surface of the Western Desert. Such areas resemble the block-strewn surfaces in the Viking lander sites (Fig. 16). Two block fields in the southwestern part of the Egyptian desert were studied and compared to basaltic hills with angular blocks 20–40 cm in diameter. In this field blocks are equidimensional, angular, with planar fractures, and slightly embedded in the sandy substrate, with few fillets, moats or wind tails. The other field consists of blocks, 20–50 cm across, which appear to have been placed there by floods during infrequent fluvial episodes. In this field the rocks are elongate, subangular, and usually with bimodal surface texture. The two block fields correlate well with the Viking lander sites in overall rock elongation form, rock

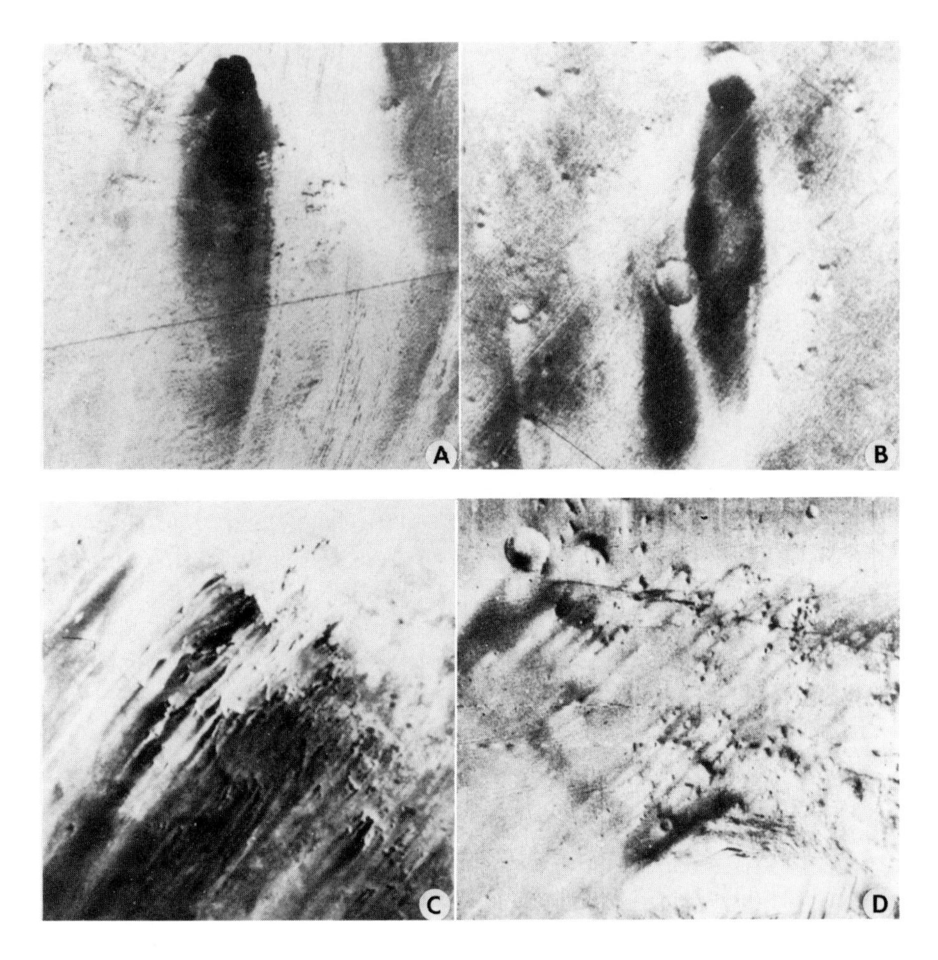

Fig. 15. (Top) Spindle-shaped, sand-free zones in the lee of a hill in the Libyan Desert (a) and in the lee of a crater in the Cerberus region of Mars (b). (Bottom) Light and dark streaks in the lee of knobs in southwestern Egypt (c) and similarly rugged terrain on Mars (d).

roundness, and the presence of facets. From this correlation, it is plausible that the blocks in the Martian sites are made of massive basalt like those in the Western Desert. Also the blocks in both cases appear to have been emplaced by cataclysmic events such as occasional floods (Egypt) or meteorite impacts (Mars).

The blocks in the Egyptian desert discussed above, although internally homogenous and non-vesicular, exhibit pitted and fluted surfaces. The pits occur singly or in rows. Often a row of elongated pits forms a flute, and some of them are pitted [20]. Wind tunnel studies of air flow over and around non-streamlined hand specimens show that windward abrasion coupled with negative air flow,

Fig. 16. (Top) A mixture of angular blocks of varying sizes on the desert surface in the Gilf Kebir region of southwestern Egypt (a) and in the Viking Lander 2 site on Mars (b). (Bottom) Wind-pitted quartzite rocks in the Gilf Kebir (c), which are similar in appearance to rocks in the Viking Lander 2 site (d).

secondary flow, and vorticity in a unidirectional wind explain the complex array of these features [20]. The pits and flutes bear a striking resemblance to similar features in Martian rocks photographed by the Viking landers (Fig. 16). This suggests that the blocks in the lander sites may have also been pitted and fluted by the wind.

As discussed above, aeolian sands in the Western Desert of Egypt, like those in many other terrestrial deserts display a thin coating on the surface, which imparts a reddish color. Whether in sand sheets or sand dunes, it was established that quartz grains are coated by kaolinite platelets, which in turn are coated by submicroscopic, powdery hematite [13]. It was further established that the coatings vary in thickness from 0.5 micrometers to about 5 micrometers; the farther the sand from the source, the thicker the coating. These coatings clearly influence the spectral reflectance of the Western Desert sands as measured from space. Considering the reddish color of the surface of Mars, it is also likely that mineral grains are covered with similarly complex coatings that affect the spectral reflectance of the Martian surface.

The above summary of correlations indicates that the Western Desert of Egypt probably exhibits the largest number of terrestrial analogies to the surface of Mars. This is perhaps due to the fact that although fluvial action played a major role in terrain sculpture, water-induced erosion did not overshadow the effects of the wind during much of the Quaternary. Furthermore, similar to what occurred in the eastern Sahara, alternations of wet and dry cycles may have also been responsible for the features on Mars.

References

1. El-Baz F: Space age developments in desert research. Episodes News Magazine, International Union of Geological Sciences (4):12–17, 1980.
2. McGinnies WG, Goldman BJ, Paylore P (eds): Deserts of the world: an appraisal of research into their physical and biological environments. Tucson: University of Arizona Press, 1968, 788 pp.
3. McKee ED, Breed CS, Fryberger SG: Desert sand sea. In: Skylab explores the earth. Geol. Surv. Prof. Pap. 1052, NASA SP-380, 1979, pp 5–47.
4. El-Baz F: Astronaut observations from the Apollo-Soyuz mission, Washington, D.C.: Smithsonian Institution Press, 1977, 400 pp.
5. Mainguet M, Callot Y: L'erg de Fachi-Bilma, Chad-Niger. Paris: Centre Nat. Res. Sci., 1978.
6. McKee ED: A study of gloval sand seas. Geol. Surv. Prof. Pap. 1052, Washington, D.C.: U.S. Government Printing Office, 1979.
7. El-Baz F, Boulos L, Breed CS, Dardir A, Dowidar H, El-Etr H, Embabi M, Grolier M, Haynes V, Ibrahim M, Issawi D, Maxwell T, McGauley J, McHugh W, Moustafa A, Yousif M: Journey to the Gilf Kebir and Uweinat, Southwest Egypt, 1979. Geogr. J., v. 146 part 1, pp 51–93, 1980.
8. Slezak MH, El-Baz F: Temporal changes as depicted on orbital photographs of arid regions in North Africa. In: El-Baz F, Warner DM (eds), Apollo-Soyuz Test Project Summary Science Report, v II: Earth observations and photography, Washington, D.C.: National Aeronautics and Space Administration, NASA SP-412, 1979, pp 7–272.
9. Logan RF: The central Namib desert. Washington, D.C.: National Academy of Sciences, National Research Council, Publication 785, 1960.
10. El-Baz F: The meaning of desert color in earth orbital photographs. Photogram. Eng. Rem. Sens. (44):69–75, 1978.
11. Walker TR: Formation of red beds in modern and ancient deserts. Geol. Soc. Amer. Bull. (78):353–368, 1967.
12. Potter TR, Rossman JR: Desert varnish: the importance of clay minerals. Science (196):1446–1448, 1977.
13. El-Baz F, Prestel DJ: Desert varnish on sand grains from the Western Desert of Egypt: Importance of the clay component and implications to Mars. In: Lunar and Planetary Science XI, 1980, Houston, Texas: Lunar and Planetary Institute, pp 254–256, 1980.

14. McCay D, Constantopolous C, Prestel DJ, El-Baz F: Thickness of coatings on quartz grains from the Great Sand Sea, Egypt. In: Reports of Planetary Geology Program – 1980, Washington, D.C.: National Aeonautics and Space Administration, NASA TM-82385, 1980, pp 304–306.
15. El-Baz F, Ondrejka RF: Earth orbital photography by the Large Format Camera. In: Proceedings of the Twelfth International Symposium on Remote Sensing of Environment, 1978, Ann Arbor, Michigan: Environmental Research Institute of Michigan, 1978, pp 703–718.
16. Breed CS, Embabi NS, El-Etr H, Grolier MJ: Wind deposits in the Western Desert. Geogr. J. (146):88–90, 1980.
17. Prospero JM, Carson TN: Vertical and areal distribution of Saharan dust over the western equatorial north Atlantic Ocean. J. Geophys. Res. (77):5255–5265, 1972.
18. El-Baz F: Circular feature among dunes of the Great Sand Sea. Science (213):439–440, 1981.
19. Whitney MI: The role of vorticity in developing lineation by wind erosion. Geol. Soc. Amer. Bull. (89):1–18, 1978.
20. McCauley JF, Breed CS, El-Baz F, Whitney MI, Grolier MJ, Ward AW: Pitted and fluted rocks in the Western Desert of Egypt: Viking comparisons. J. Geophys. Res., v. 84, n. B-14, pp 8222–8232, 1979.
21. Gifford AW, Warner DM, El-Baz F: Orbital observations of sand distribution in the Western Desert of Egypt. In: El-Baz F, Warner DM (eds), Apollo-Soyuz Test Project Summary Science Report, v. II: Earth Observations and Photography, Washington, D.C.: National Aeronautics and Space Administration, NASA SP-412, 1979, pp 219–236.
22. Strain PL, El-Baz F: Sand distribution in the Kharga depression of Egypt: Observations from Landsat images. International Symposium on Remote Sensing of Environment, 1982. Ann Arbor, Michigan: Environmental Research Institute of Michigan, First Thematic Conference: 'Remote Sensing of Arid and Semi-arid Lands', Cairo, Egypt, ERIM, paper B-11, pp 101–102, 1982.
23. Bagnold RA: The physics of blown sand and desert dunes. London: Methuen and Co. Ltd., 1941, 266 p.
24. Embabi NS: Barchan dune movement and its effects on economic development at the Kharga depression, Western Desert, Egypt. Cairo, Egypt: Ain Shams University, Bull. Mid East Res. Ctr., 1978 (in Arabic).
25. Goetz AFH: Reflectance radiometry of the earth: preliminary results from the Shuttle multispectral radiometer. In: Lunar and Planetary Science XIII, 1982, Houston, Texas: Lunar and Planetary Institute, pt 1, pp 267–268, 1982.
26. Beadnell HJL: An Egyptian oasis: an account of the oasis of Kharga in the Libyan Desert. London: Murray, 1909, 248 p.
27. El-Baz F, Maxwell TA: Eolian streaks in southwestern Egypt and similar features on Mars. In: Proceedings of the Tenth Lunar and Planetary Science Conf., Houston, Texas: Lunar and Planetary Institute, Pergamon Press (New York), pp 3017–3030, 1979.
28. Manent LS, El-Baz F: Effects of topography on dune orientation in the Farafra Region, Western Desert of Egypt, and implications to Mars. In: Reports of Planetary Geology Program – 1980. Washington, D.C.: National Aeronautics and Space Administration, NASA TM-82385, pp 298–300, 1980.
29. Bagnold RA: Libyan Sands. London: Hodder and Stoughton, 1935.
30. El-Baz F, Breed CS, Crolier MJ, McCauley JF: Eolian features in the Western Desert of Egypt and some applications to Mars. J. Geophys. Res. (84) n. B-14:8205–8221, 1979.

Author's address:
Itek Optical Systems,
10 Maguire Road,
Lexington, MA 02173, USA

2. A classification of dunes based on aeolian dynamics and the sand budget

Monique Mainguet

Introduction

Basic definitions

A dune is usually defined as an accumulation of loose particles, deposited or reworked by the wind, with diameters varying from two or three milimeters to tens of micrometers. In fluvial environments (in seas, lakes or river streams) one can also find accumulations of loose particles deposited or reworked by the water, which are analogous to aeolian edifices. However, such banks of clay, sand ridges, and under water ripple marks, up to 30 m in height, cannot be considered as dunes because of the lack of sharp crests.

Dunes are subdivided into three categories according to decreasing amounts of activity: active dunes, fixed dunes and vegetated dunes.

The term active dune is reserved for constructions of aeolian material independent of the size of the particles (there are dunes of argile, of lime, of sand, or of mixtures of two or three of these), or of the nature of the grains.

The composition of dune sand is dependant upon the geographical location. In the deserts of tropical latitudes, where the majority of the material is furnished by coats of the weathered rocks from the tropical, humid climates, the sand is essentially siliceous. In cold Antarctic deserts, dunes are mostly composed of grains of ice or snow. At middle latitudes dunes are usually composed of salt grains, as in the gypsum deserts in the state of New Mexico; in the littoral areas of the mediterranean they are composed of organogenic materials like shell sands; in the volcanic regions, as those of the San Francisco volcanic field around Flagstaff, Arizona and on the Island of Hawaii, the dunes are composed primarily of ash.

A sandy edifice loses its characteristic activity as soon as pedogenetic mechanisms intervene or when the dune sand is highly cemented by iron oxides or encrusted by gypsum, carbonates, iron or even silicon. The second mechanism of fixation is the result of winnowing and the formation of a pavement of coarse sand particles with dimensions exceeding those exportable by the wind. The third

El-Baz, F. (ed.), Deserts and arid lands. ISBN: 90-247-2850-9.

cause of dune fixation is stabilization by vegetation cover by varying types and densities: by a steppe of small grasses, a steppe of grassy bushes or trees, by a savanna, a woody savanna or by tropical or evergreen forests. It is in such areas that the process of reactivation of the vegetated dunes become significant when climatic changes occur.

Classifications of dunes

There have been numerous studies dedicated to the classification of aeolian edifices. Of all of these, the most traditional have been topological, where the dunes were classified based on form and disposition [1–32]. Most of such classifications have failed to sufficiently address the genetic parameters within the topological studies. In this paper a classification of dunes based on dynamic criteria is proposed. It explains the varying aspects of dunes and differentiates between the topological form, the direction of winds, and evolution in time. It also takes into account the budget of sand, the difference between the sand which is imported and exported by wind transportation, in the same region.

In a region where the importation of sand surpasses the possibilities of sand exportation, active sandy edifices are formed called depositional dunes. Their volume and number increase as long as the feeding of sand is in excess. When in an area of abundant sand, exportation becomes the dominant process, sandy edifices appear, for which the proposed name is erosional dunes. This name is selected because of the characteristic form and direction, which result from deflation, i.e. from erosion and aeolian exportation.

In 1931, L. Aufrere [3] proposed a dynamic classification of dunes of his 'morphological cycle of dunes', in which he distinguished amongst:

Winds in conjunction: This occurs when the dominant directions diverge, scarcely distant from one another. This is the regime of trade winds that S.G. Fryberger [33] called 'narrow and wide unimodal'. The distinction between a narrow and a unimodal direction is more than a narrow 'fork' of wind. Moreover, it is an oscillation of winds forming a small angle from one or more specific direction. It can also include many wind directions where only one is effective.

Winds in opposition: This occurs where the winds have two dominant directions at an angle of about 180°. Examples of this include the regions of Chergui and Saheli of meridial Moroco, of the Alizé and the Sirocco in south Algeria and Tunisia as well as the monsoon and trade winds of the Sahara.

Incident winds: These are bidirectional regimes that are diverse and are subdivided into numerous branches by topographical irregularities and thus cannot be discerned by a network of anemometers. This is the 'acute and obtuse bimodal' regime described by Fryberger [33].

Multidirectional winds: These are the aeolian regimes with at least two dominant winds that create a 'complex' regime [33].

Thus, the active dunes can be classified into three families corresponding to three principal aeolian regimes: (a) crescent dunes and transverse chains – response to an anemometrical regime that is dominated by a monodirectional wind; (b) linear dunes, the result of an anemometrical regimes of wind that is continually disrupted by topographic irregularities or bidirectional winds, and (c) star dunes, the creation of an anemometrical regime without a dominant wind direction.

Although this classification was given by several authors, [3, 4, 11, 33], it is not enough to classify the dunes based only on their dynamics. It is necessary to consider two other fundamental parameters, the budget of sand and the directions of accumulation and exportation. Thus, it is here proposed to distinguish between: (a) active depositional dunes including barchanic edifices and sandridges, linear dunes, and star dunes; and (b) erosional dunes including parabolic edifices, and sandridges.

Aeolian features can be classified as isolated edifices of groupings of dunes, in which one is able to distinguish their organization, and forms that are separated by open ground. In each subclass of dunes distinctions can be made between texture or form, and structure or pattern.

In the classification of Breed and Grow [12] distinction was made between (a) simple (or elementary) dunes, the most fundamental forms, (b) composite dunes, which are comprised of the same fundamental elements of the simple type but they occur in greater numbers or in different sizes. The composite forms are associations of individual components which can be of equal or unequal dimensions; and (c) complex dunes, which are composed of many varieties of different dunes, combined by juxtaposition or superposition.

The principal dune patterns are: (a) isolated dunes where the placement of each structure is independent of its neighbors and the alignment of the dunes is based upon a preferential direction; and (b) groupings of dunes where alignment is in many directions, and in triangular and checkerboard patterns.

Active depositional dunes

Crescentic and transverse chains

This class of dunes is characterized by the importation of sand to the dune in a direction perpendicular to the largest dimension of the edifice. Here we can distinguish between two categories: dunes growing in barchanic crescents, and barchanic chains formed by coalescing crescents. The form and the individuality of crescents can be more or less precise; the transverse chains occur when the crescents have been erased.

Crescent dunes. The crescent dunes have three common paramaters: (a) the crescent shape (b) an axis of symmetry, which conforms to the dominant direction of the wind responsible for forming the edifice, or the middle vector of the annual wind; and (c) two topographically different faces, separated by a crest line that rigorously delimits the two sides.

Isolated crescentic dunes exist in regions where transport is dominant, where there is no topographic blockage or other natural obstacles (vegetation, positive or negative relief), or any man made obstacles (village, or palm tree plantation).

Crescent dunes are organized in linear chains that form at the same time as the deceleration in the movement of the barchans, without a decrease in sand feeding. As constituents of numerous ergs, they are separated by interdunal spaces where sand is rare, or even absent.

As long as sand is imported, the edifices, or transverse barchanic chains tend to climb one on top of another into very densely packed areas, where the interdunal spaces become increasingly smaller, until they no longer exist, thus resulting in *akles*, which are described in the French literature [34].

The *akles* are very dense assemblages in which the sand is highly concentrated, but from which the sand can escape while migrating from one ridge to another in a direction transverse to the ridges exporting the sand, to perhaps outside of the arid zone entirely. This is possibly the case in the south of the Sahara, where drought has caused an increase in the movement.

The dynamics of crescent dunes are dependent upon the dimensions of the edifice. The isolated barchans can be considered vehicles for the transport of sand. The direction of crescent dune migration conforms to its axis of symmetry.

Barchans. A barchan is a dune which, in projection, has a crescentic form, convex to the wind, with two face meeting at a sharp angle (Fig. 1). The barchan is thus an edifice that is transverse to the wind, with oblique horns. Its dimensions range from several tens of centimeters to more than one kilometer. The back, or reverse of the barchan faces the wind and is a slope of landing and migrating sand, attaining angles between 6° and 12°. The front is a leeward slipface with a streeper slope, between 20° and 33°. This also serves as a slope where sand can arrive, either carried by forward or reverse vortices. In this instance, it assumes the role of an active slope swept by vortices, the axes of which are normal to the surface of the slipface. Lobate sand slides then appear.

A barchan is a dune that migrates as a whole. The rate of migration is dependent upon the size and the slope on which the edifice is moving. The speed decreases when the size of the edifice grows. As long as the surface of the ground is inclined in the direction of the wind, the rate of progression of the sand accelerates; if it is against the slope, the rate decreases [35].

Symmetrical barchans are exceptional. Sometimes, only one horn will subsist, growing into a strongly accentuated 'dexter' or 'senester' horn, which eventually

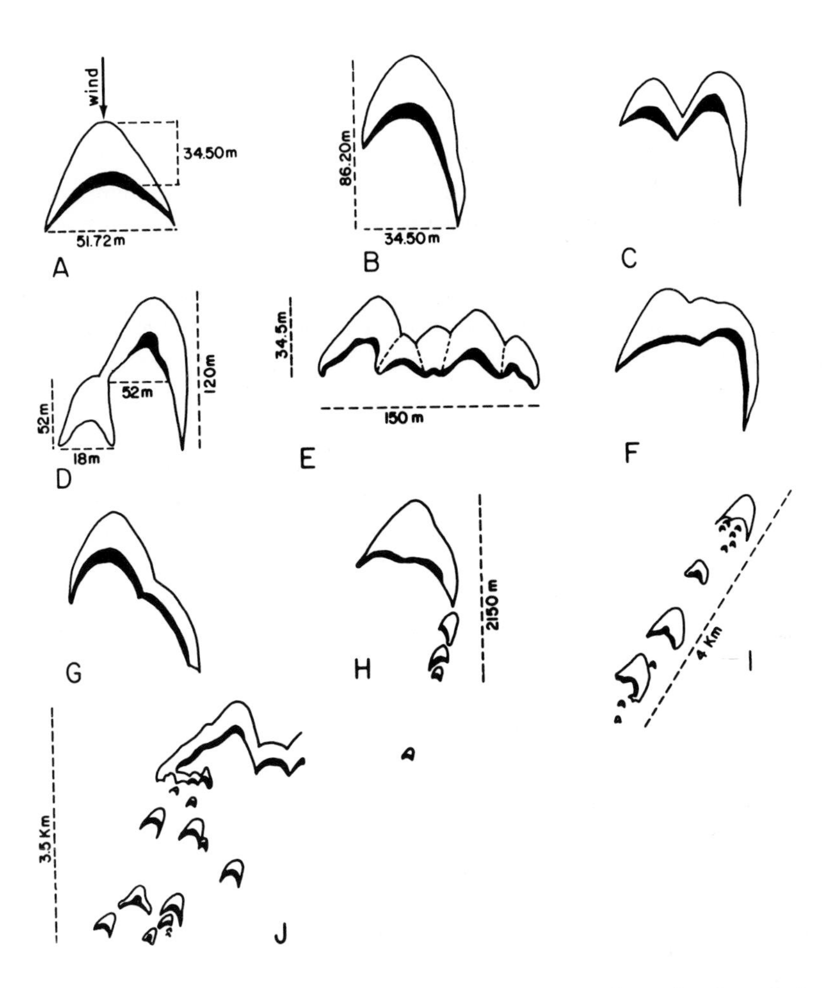

Fig. 1. Barchans: (a) simple barchan; (b) asymmetrical barchan; (c) coalescent barchans: the horns are joined without the alteration of the crescent shape; (d) coalescent joined barchans; (e) barchanic chains with visible crescents; (f) barchanic edifices in which the central barchan has been erased; (g) asymmetrical barchan, the elongation of the right horn results from the coalescence of a second barchan; (h) large barchan with one horn nurturing four smaller barchans; (i) convoy of barchans with smaller barchans leewardly embrased by larger barchans, disproving the hypothesis of negative flow in the leeward area of a barchan; (j) ensemble of barchanic figures with decreasing size in the downwind direction, indicating that the upwind barchan traps the sand. The smaller barchans submit to this 'shadow effect'.

replaces the barchan by a linear dune.

The necessary aeolian regime for the formation of a barchan is a dominant monodirectional wind. When a barchan arrives on the thick, sandy margin of an erg, it melts in the sand sheet and disperses. This explains how the barchans can only exist while they remain on the border of an erg; these dunes are not constituents of ergs, but rather they serve as vehicles of sand transport into or out of the ergs.

Prebarchanic forms. In 1923, Hogbom [36, 37] assigned the term 'shield' to barchanic edifices with a certain positioning of their concavity, enabling them to be classified into barchanic shields, dihedral barchanic edifices, and barchans (Fig. 2).

The shield is the least elaborate barchanic form (cake-like sand dune). It is a result of sand deposition in an oval shape, with a convex profile and an oval form, able to attain many meters in height. Oldham [38] pointed out that crescentic dunes usually result from a transformation of the shield-like shape. The shield is formed without topographic obstacles in an area where the wind is curved upward. The sandy accumulation is thus formed, curving the upward air flows, manipulating the sand in a lateral direction and causing the dune to curve upon itself so it will not attain excessive elevation. Thus, it assumes the form of an arc of a circle.

The shield type edifice remains in its shape as long as it can conserve a modest height, and its slope remains gentle. When the shield grows, its air currents are raised, vortices form and the barchanic cut appears; the shield shape then begins to evolve into the barchanic shield shape.

The barchanic shield (juvenile crescentic sand dune) is an active dune with a concave cut on its leeward face. The longitudinal and transverse profiles of the dune remain convex. When viewed from above, its downwind side appears gouged, but it does not form a sharp crest, for this downwind cutting at a feeble angle does not function as a slip face.

The dihedral barchan (crescentic sand dune) cannot be considered as a shield dune because the passage of the back of the dune towards the front creates an angular crest line which results from the cutting of the downwind slope. This hinders the development of a dune summit. Until the barchanic concavity reaches the summit of the edifices it cannot be considered a barchan. In the dihedral barchan, the back side of the dune is larger than the upwind side of the edifice. This edifice also evolves by the avalanching of the sand on the front face.

The position of the concavity of the shield is not fixed, for a sand storm can easily modify it. That of the dihedral barchan, however, is stable. Nonetheless, the principal difference between a shield with a barchanic cutting, a dihedral barchan, and a barchan, is in the position of the concavity in relation to the oval shape. In the latter two cases, the crest is angular. It is located at the summit of the dune of a barchan, and somewhere in the leeward slope of a dihedral barchan. Theories of wind circulation about a barchan are illustrated in Figure 3.

A barchan grows and maintains itself on a planar surface in a monodirectional aeolian regime where the wind direction scarcely varies. All topographical or aerodynamical obstacles which create vortices are detrimental to the dune. As

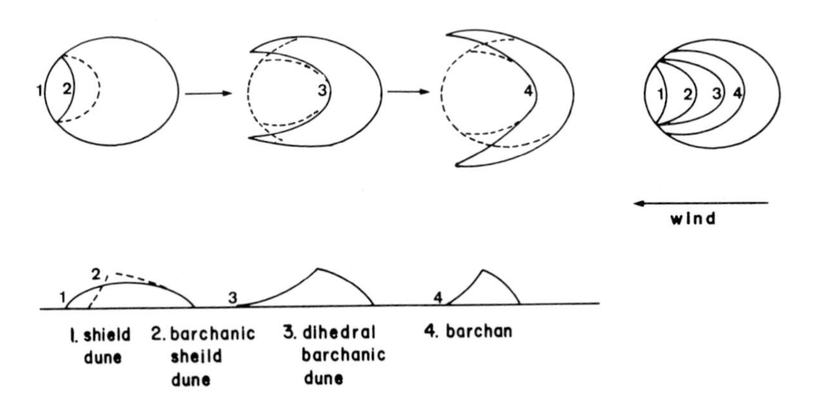

Fig. 2. The prebarchanic shapes of dunes; shield dune; dihedral barchanic dune; barchan.

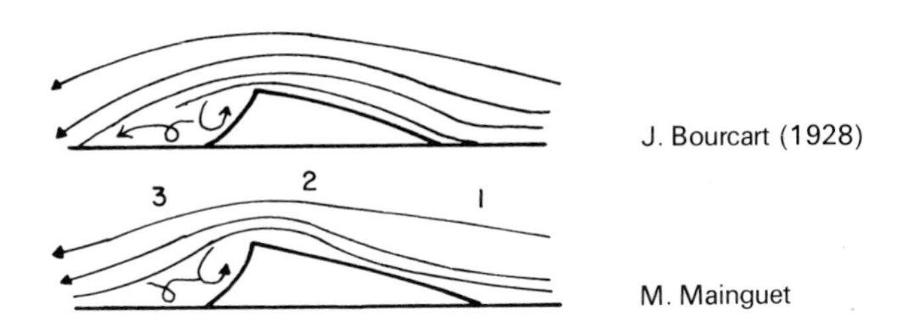

Fig. 3. Dynamic aeolian activity around a barchan, after [11].

soon as a barchan encounters an obstacle, it deforms and transforms itself into an entirely different structure (e.g. into a sand arrow). After the sand transport has surpassed the obstacle, the barchans tend to reform. Also, an obstacle does not cause a problem in barchanic conservation, except when it causes the wind currents to divide.

Composite barchanoid forms occur in numerous varieties (Figs. 1 and 4) including: giant barchans formed of small jointed barchans; giant barchans, the backs of which are fashioned into smaller barchans; chains of crescents in shaded forms (lines of erased crescents); and belts of barchans, normal to the wind (crescentic sand chains).

The patterns of barchanic belts include chains of coalescent crescents (Fig. 4) and groups of triangular barchans in a form resembling the shape assumed by a swarm of migrating geese (Fig. 5).

This disposition of barchans in 'barchan swarms' can be observed at the southern foot of the Tibesti Massif in Chad [39] and to the east of the Gilf

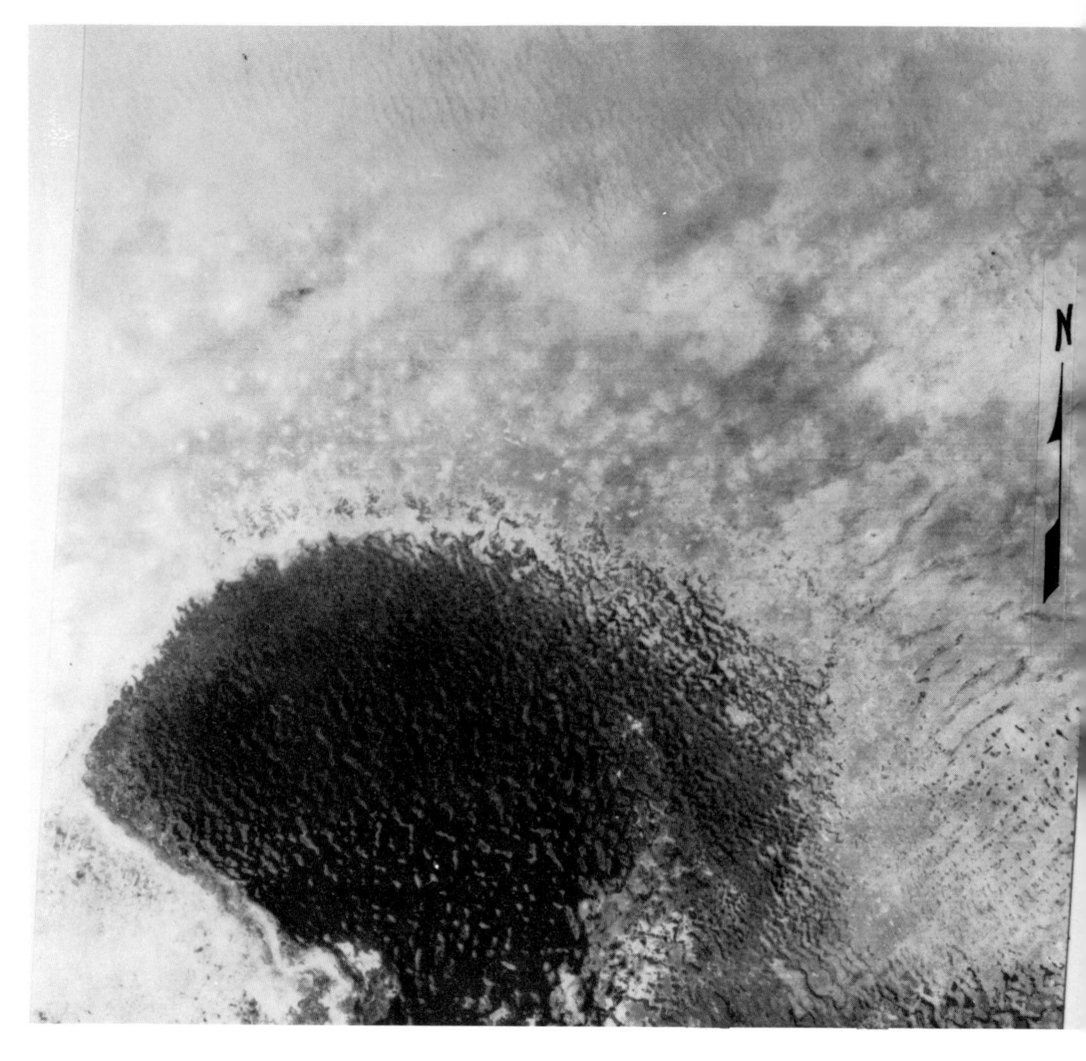

Fig. 4. Transverse dunes upwind of Lake Chad. Perpendicular to these transverse edifices appear longitudinal sandridges and depressions.

Kebir plateau in Egypt. Such collective forms are often preceded by a main dune; a large upwind barchan followed in the direction of the wind by smaller barchans.

The alignment of isolated barchans can take the shape of trains of barchans oblique to the wind or of periodical chains transverse to the wind. This pattern leads to transverse chains (Fig. 6). Barchans and all the barchanic forms as well as barchanoid chains are transverse aeolian edifices. The existence of a downwind slip-face constitutes the point held in common by all barchanic forms.

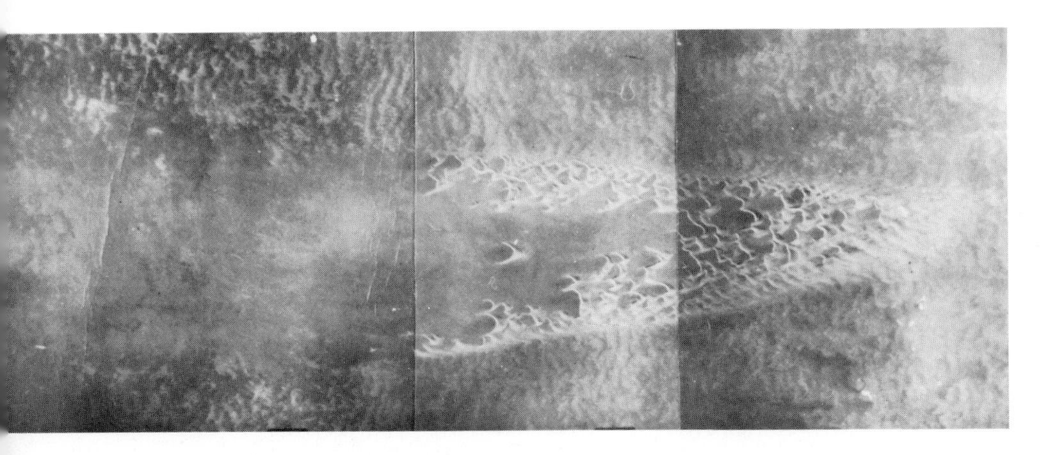

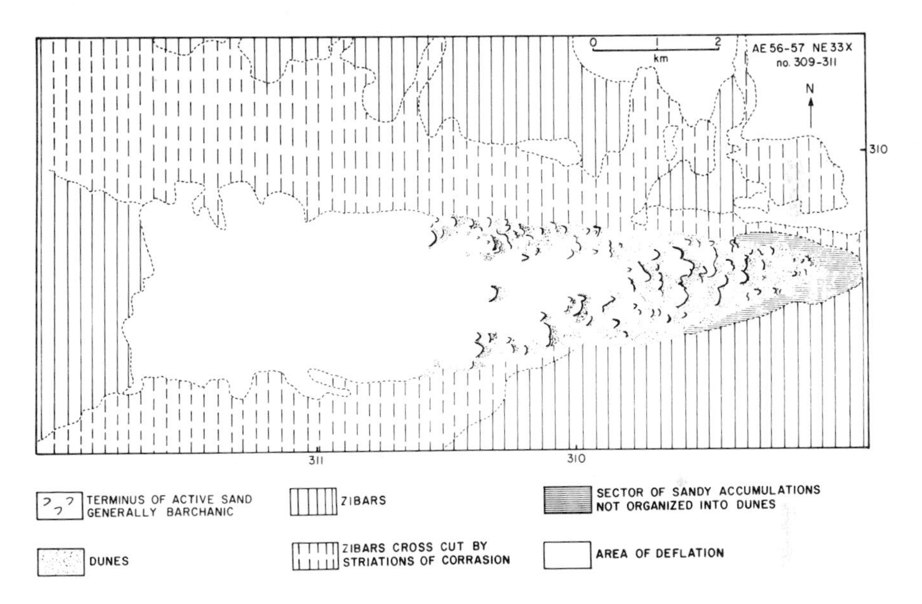

Fig. 5. Barchanic swarms in the pattern assumed by migrating geese; (top) mosaic of three aerial photographs; and (bottom) drawing of same area.

Linear or elongate dunes

The term 'longitudinal' which had been used by Bagnold [40] is considered inappropriate here. This is due to the genetic connotation that many authors assign to it, giving it a sence of 'conforming to the direction of the dominant winds'. The terms linear or elongate, however, do not imply a genetic connotation.

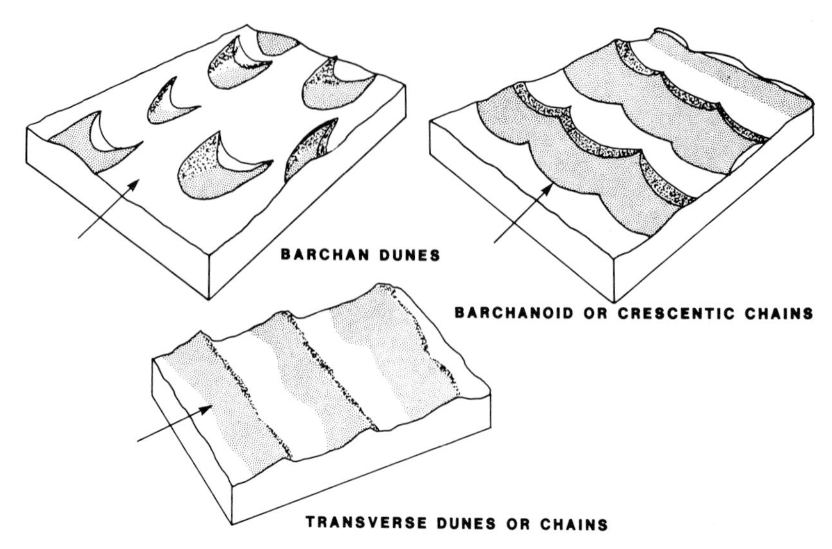

Fig. 6. Transverse dunes or chains [69].

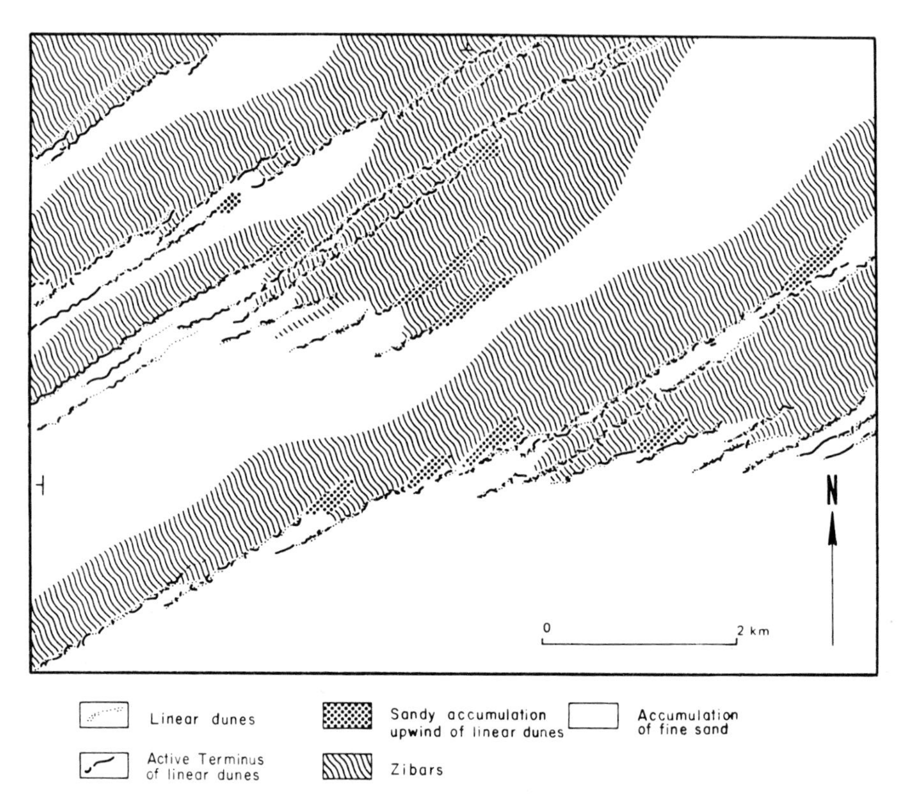

Fig. 7. Linear dunes without an obstacle combined with 'zibars', giant ripple marks in the Erg of Fachi Bilma.

Simple linear dunes. A linear dune is a sandy edifice, narrowly elongate at the base, placed in a rectilinear direction, and sinuous in detail (Fig. 7). It is especially elongate in one direction with a profile formed by two steep slip faces that meet in an angular crest.

The relationship between the length and the width show that the length of a linear dune is always many times (up to ten) greater than its width [39]. For example, the longest linear dune of the erg of Fachi Bilma is 40 km in length and varies in width from 50 to 100 m downwind, with an average length to width ratio of 80.

The average linear dunes have lengths from 2–3 km and widths of 30–150 m. Furthermore, they are often arranged in staggered figures, in systems that are sufficiently elongate as to attain 30–40 km in length. In this pattern, the linear dunes succeed each other with only brief interruptions.

Bagnold [40] explained the genesis of linear dunes as being the elongation of a barchanic horn. He pointed out that a linear dune is an obstacular dune and must also conform to secondary winds. However, this interpretation does not account for the following: (a) linear dunes exist that are neither anchored on a barchan nor on any topographic obstacle (Fig. 8); (b) linear dunes of varying directions occur in the same sector leeward of buttes of different forms. The angular differences between these dunes constitute the response to the different deviations of the wind around the different forms of the buttes; and (c) when an aeolian current encounters an obstacle, a triangular area with high vorticity forms leeward of the obstacle. The transported sand is deposited tangentially to the limits of this area. Thus, a linear dune forms obliquely to the left or to the right of the dominant wind direction.

Three preferential locations of linear dunes are considered here: (a) downwind of a fixed rocky obstacle. The linear dune is a dune of 'sillage' (this is a French word expressing a phenomenon similar to the bow waves in the wake of a boat), formed by the deposition of sand tangentially to the wind vortices created downwind of the obstacle; (b) downwind of a mobile edifice, formed, for instance, by a growing branch of a barchan; and (c) dunes without an apparent anchoring obstacle or sometimes in an arrangement of *zibars* or sandridges.

Linear dunes, being oblique dunes of 'sillage' are dexter if oriented to the right of an obstacle in their relationship to the wind direction, and senester if oriented to the left of the wind. In the sand sea of Fachi Bilma, it has been demonstrated that, downwind of the Tibesti Massif, at the begining of the erg, the linear dunes are of a senester deviation. The same is true upwind of the Air Massif, against which the active erg of Fachi Bilma is stopped. The linear dunes at the heart of the erg are dexters, as in most of the north-equatorial ergs.

Active linear dunes appear in rocky areas where irregular topography causes the aeolian flow to lose its characteristic unidirectional regime and divide around topographic obstacles. Linear dunes are also formed in regions where the aeolian

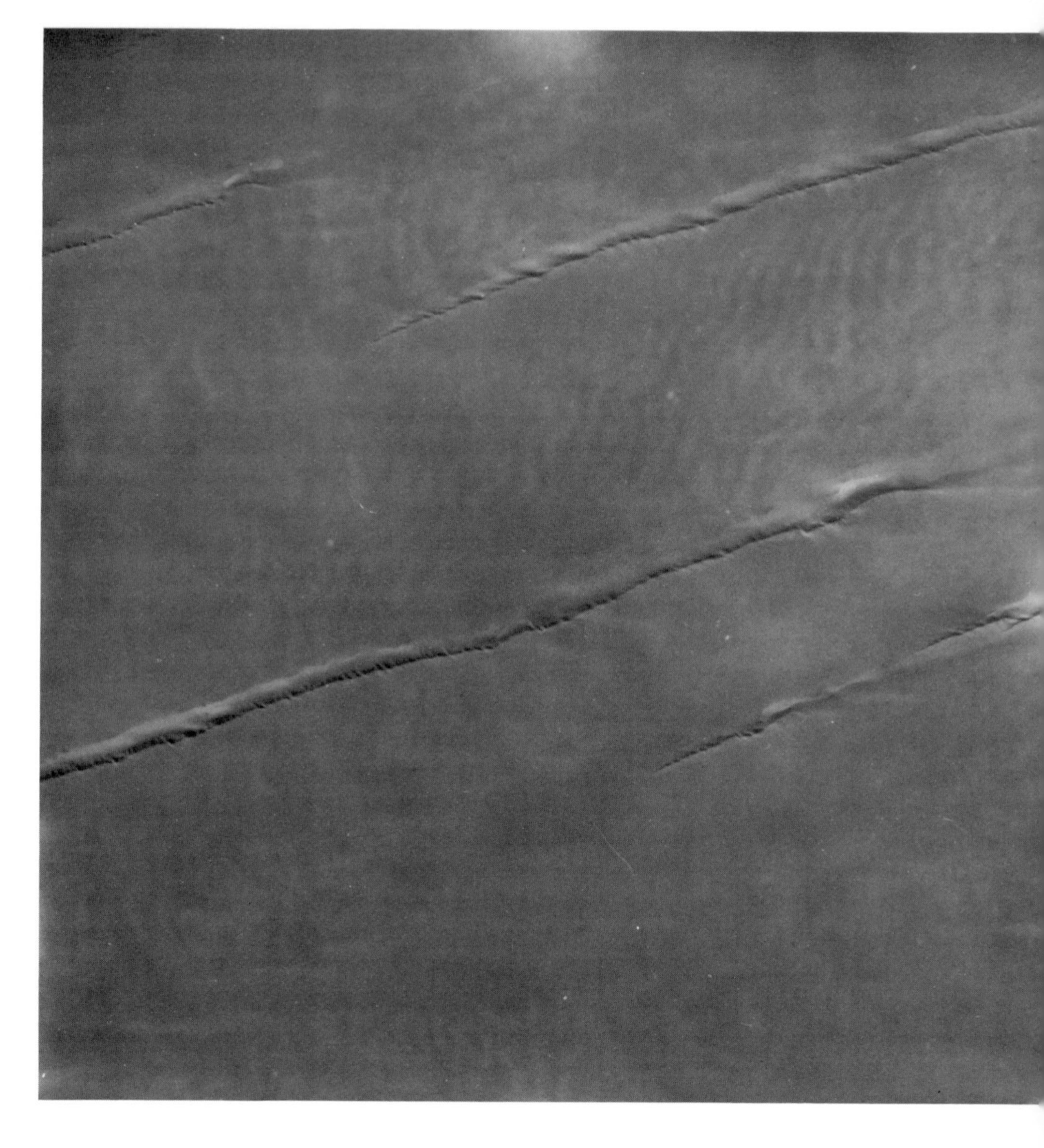

Fig. 8. Aerial photograph of the erg of Fachi Bilma (18° N, 15° E) showing linear dunes that originate without an obstacle. Some of the dunes display a small mound at the upwind origin.

regime is of two dominant orientations. A linear dune formed tangentially to the 'sillage' of any obstacle whatsoever will be oblique in its relationship to the annual resultant wind; it is asymmetrical, with one side of the wind being a slope of deflation, and the other a slipface. On the downwind face, vortices form, returning the sand to the dune.

During field investigations in December 1980 in the Oubari erg in the north of Sebba, Libya, a sand storm was observed, supporting the results of Tsoar [41], which include: (a) the sand displaced by the wind arrived at the dune where a part of the sand was already fixed on the upwind side; (b) the other part crossed the crest of the dune, and, on the downwind side was lifted by the effect of the vortices of 'sillage' and then returned to the dune; and (c) the sand next to the dune finally migrated parallel to the axis of the edifice, elongating according to its own direction, resulting in an elongate dune.

Composite forms. The major forms of composite dunes are: (a) the bundles of parallel linear dunes; and (b) the feathered linear or 'bouquet' dunes. The parallel dunes are usually scattered in patterns, which are staggered either upwind or downwind (Fig. 9). In the former, the termination of a linear dune is downwind of the head of the preceding upwind one. In the former, the termination of a linear dune is downwind of the head of the preceding upwind one. In the latter case, the end of the linear dune is upwind of the head of the following one.

The feathered linears or 'bouquet' dunes can be symmetrical or asymmetrical (Fig. 10). Feathered linear dunes are combinations in which we can speak of (a) a stem side, (b) a bunch side. In the case of asymmetrical feathered dunes, they consist of a principal linear dune; (c) the longest and most rectilinear of them have the shortest linear dunes anchored to them; and (d) making angles with them that are more or less open.

The feathered groupings of linear dunes give indications to the directions of the wind. The bunch side indicates the upwind direction of the aeolian current; the stem side indicates the downwind current. In a 'bouquet' shape the principal linear dune is upward and the secondary ones are downward. The principal oblique linear dune thus appears to be a generator of vortices which are responsible for the angular differences between the secondary dunes and the directions of the wind.

Complex combinations. The most common complex combination is that of linear and crescentic dunes (Fig. 11). In this case, linear dunes can be formed along the preferential elongated barchanic horn, or combined in a figure formed by linear dunes and barchans, emitted downwind. In the erg of Fachi Bilma the difference between the axes of the barchans and those of the linear dunes creates an angle of 34° [39].

Another combination is that of linear dunes and giant ripples. Here, there is a superposition of linear dunes on top of sandridges with great variation between the angles of the two. Linear dunes are fashioned by the winnowing of the finest sand particles of the sandridges.

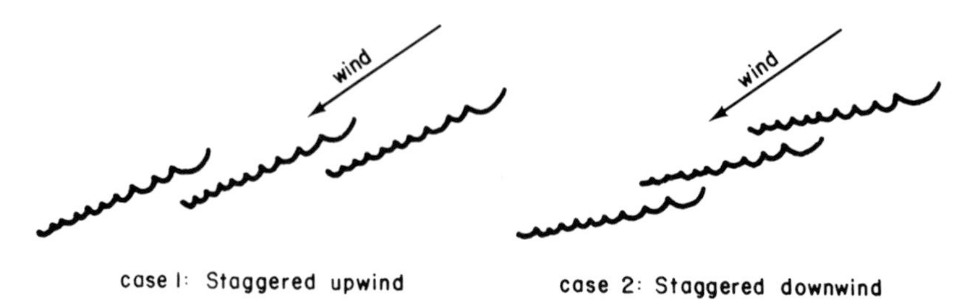

Fig. 9. Examples of staggering in the alignments of 'dexter' linear dunes, case I is seen more frequently.

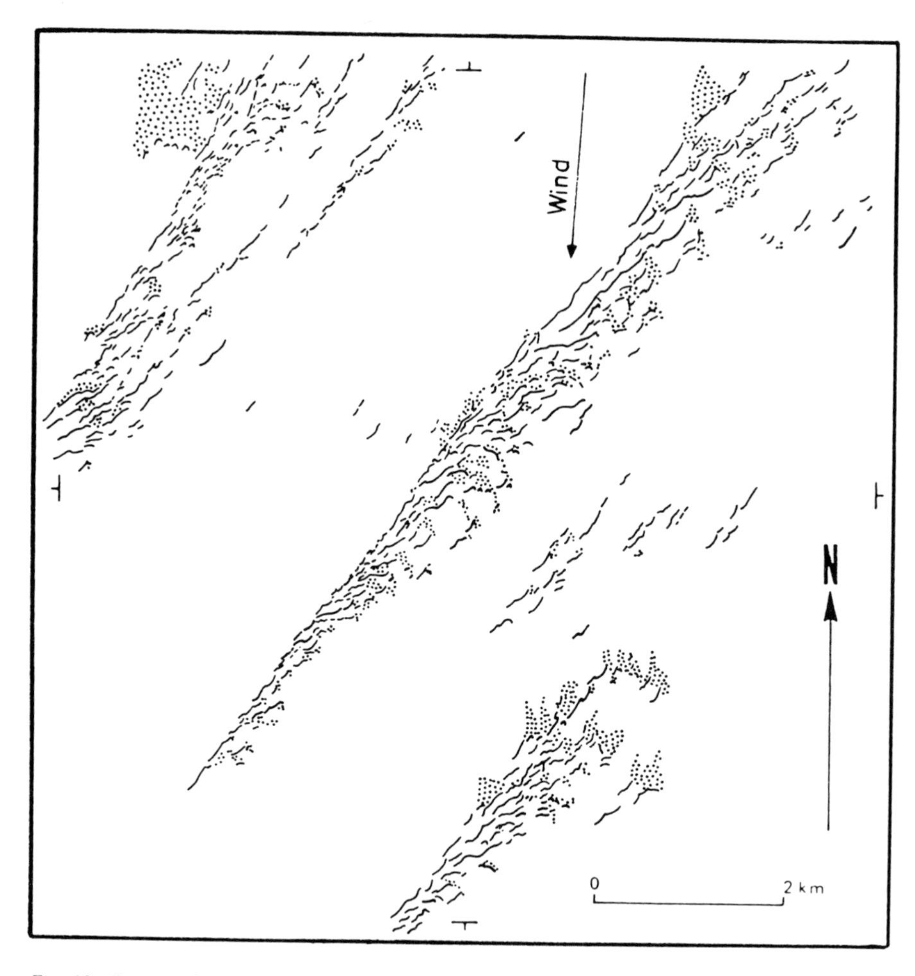

Fig. 10. 'Bouquets' of linear dunes. The central stem is nearly closed and asymmetrical in its downwind part. The southern bouquet is much less asymmetrical and more open. The bouquet always opens upwind and closes downwind; the most regular side of the bouquet is upward and the most irregular, leeward. For this reason bouquet dunes can be used as indicators of wind direction.

Fig. 11. Landsat image of Bir Nakhlai, Western Desert of Egypt, showing barchan transformation into linear dunes and swarms.

Star dunes

A star dune is a sandy hill in the form of a pyramid, from the summit of which emanate three or more arms (unless it possesses more than two, it is considered a barchanic edifice). These arms always form a dihedral profile with an active crest and a slip face. They are, in fact, linear dunes which are sinuous and disposed on a slope. A pyramidal dune may have a peaked, elevated central point, or else a less precisely peaked or shapeless one from which the arms diverge.

Star dunes reach the greatest heights of all terrestrial sandy edifice. Their height surpasses 400 m in the Taklimakan Desert of northwestern China, in Erg Issaouane and in the Dasht-e-Lut of Iran. Those of the Issaouane N'Irrarene sand sea in Algeria attain up to 350 m in height. These dunes are edifices that display very large basal diameters. A maximum diameter of 3 km is attained in the south of the Erg Oriental (from 30° N and 8° E). In China's Ala Shan Desert (40° N–100° E), the diameters are up to 1 km.

Star-shaped dunes are not very abundant. They usually appear at the borders of ergs near topographic obstacles, except in the Erg Oriental, where they are either arranged at random or else aligned in lines, constituting the majority of the sand sea. Their occurrence near topographic obstacles is relevant to the dynamic environment, which is only found in the areas where several winds encounter one another.

These star dunes may only be explained by the existence of a strong vertical component which is exaggerated when two or three winds meet. It is interesting to see that the erg of Fachi Bilma, with a length of 800 m, found at the southwest foot of Tibesti (downwind of the trade wind) begins with the encounter of two branches of the trade winds by three chains of star dunes.

In the Erg Oriental there is a transition from dunes with regularly diverging arms, to one side-armed dunes, progressively replaced by star dunes. Another complex form associates large and small arms of different lengths. One complication occurs when, in the multi-directional aeolian wind regime, a certain direction becomes slightly preponderant. This explains the forms of the star dune type with crescentic curvature which are found in the Rub Al Khali in Saudi Arabia and in Gran Desierto in Mexico, where the dunes assume the shape of a jellyfish.

Distinction can be made between the following types of star dune groupings: (a) random arrangement of isolated edifices; (b) randomly arranged groupings with coalescent arms; (c) chains of isolated edifices or of contiguous edifices; and (d) a checkerboard arrangement with triangular or diamond shaped interdunal spaces, as in the ergs at the northeast of Tassilis N'Ajjer of South Algeria.

In the southwest of the Erg Oriental, the star dunes form straight lines. This alignment of pyramidal dunes occurs when deposition is substituted by erosion.

Active erosional dunes

Parabolic dunes (Fig. 12)

A parabolic dune is a sand crescent that is concave upwind, with fairly long

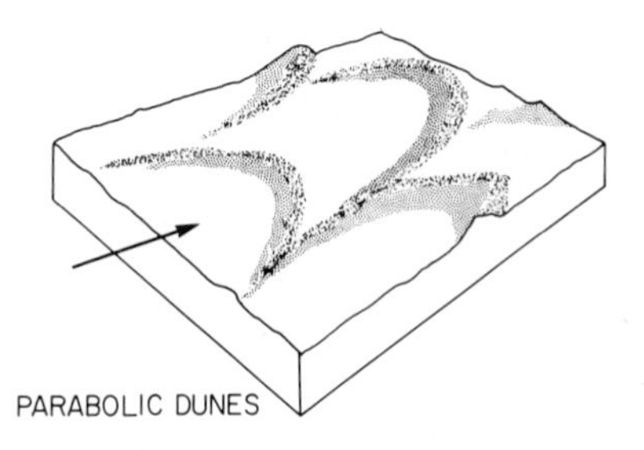

Fig. 12. Parabolic dunes [69].

horns. A transverse profile across a parabolic dune displays an upwind slope in its concavity that is greater than the downwind slope.

Numerous authors described parabolic dunes in India [11, 42], in France [43, 44], in Hungary [36, 37], in Australia [45] and in the White Sands of New Mexico [46]. This distribution shows that the dunes appear in essentially littoral zones, semi-arid or sub-humid zones. They are a transitional form between active dunes and vegetated dunes, but have the dynamic demeanor characteristic of active dunes.

In the Plains region of Texas, the mechanisms responsible for the growth of parabolic dunes are well displayed. A sand sheet which is strewn with hills of sand and also covered with dense vegetation partially reactivated. On the face of the sand hills, which are exposed upwind toward the west, 'blowups' appear in the vegetation cover. They are gradually transformed into furrows up to one meter deep, which can appear as crescentic shapes when viewed from above, but most of them are aligned along the axis of the hill, eroding it into a crescentic parabolic dune. When the body is completely cut along its axis, longitudinal dunes of erosion parallel to the upwind direction appear. This is the last phase in the evolution of parabolic dunes in the instance of non-sand-carrying winds when there is an excess of available energy.

McKee described a new sequence which ends in the development of parabolic dunes, in the White Sands of the state of New Mexico [46] (Fig. 13). The winds in this area are multidirectional, but only the winds blowing from the west or from the southwest are capable of moving sand. From the upwind to the downwind margin of the dune field, one can observe (a) dome dunes; (b) transverse dunes that are almost rectilinear, always positioned perpendicular to the efficient wind; (c) barchans; and (d) parabolic dunes in 'U' or 'V' shapes, formed by 'blowups'.

In Niger, in a region situated to the north of Zinder (Fig. 14), another type of relationship appears between barchans and parabolic dunes. Placed on a substrate that is slightly clayey are round or oval sandy edifices which are inactive. Their sand, taken *en masse* by the pedogenetic action of a paleoclimate, is covered by a grassy steppe, as long as the dunes are not cultivated.

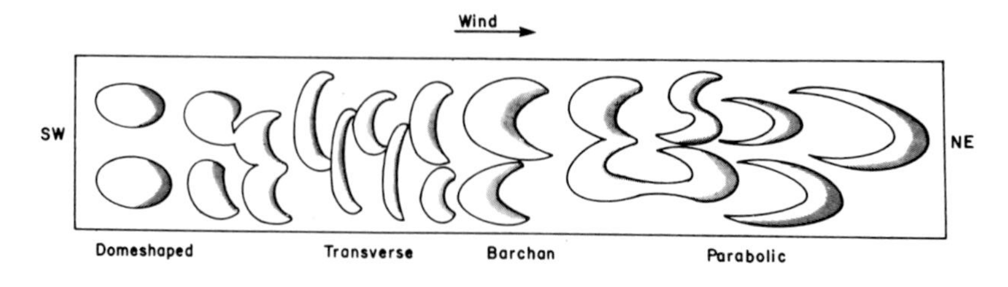

Fig. 13. Evolution of dune types at White Sands [46].

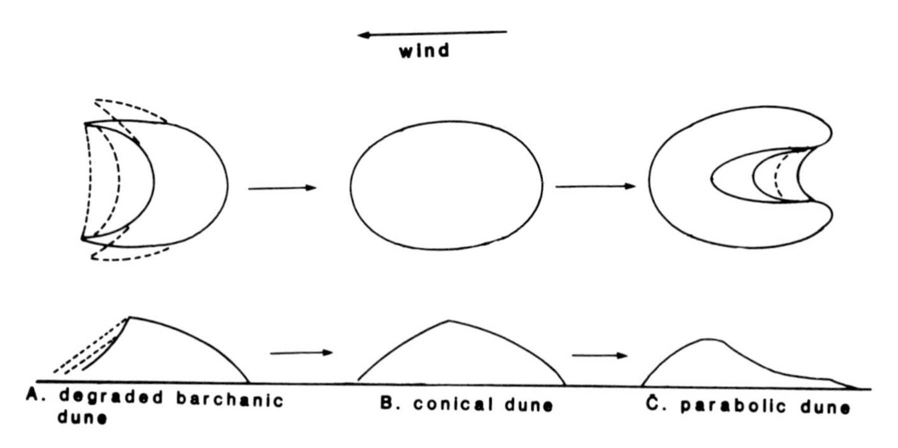

Fig. 14. Scheme of the evolution of crescent dunes in the north of Zinder (Niger): (A) degraded barchanic dune, (B) conical dune, (C) parabolic dune.

The examination of these edifices reveals three types of dunes: (a) barchanic edifices in concave crescentic shapes facing towards the southwest that are degraded and flattened; (b) conical edifices; and (c) parabolic edifices oriented towards the northwest, which occur in perfect conformity with the northeast-southwest direction of the trade winds.

It is believed that these parabolic dunes constitute the final phase in the evolution of immobile barchans. The latter in this case are taken *en masse* and sculpted by corrasion into parabolic dunes, with passage from an arid to a sub-humid climate (800–1000 mm precipation) to a quite dry one resembling the present climate.

Simple parabolic dunes can be divided into: (a) hairpin shaped dunes, e.g. White Sands, New Mexico, USA, and dunes of the region of Gascony, France; and (b) V-shaped dunes, e.g. the basin of the Columbia River, Washington State, USA. The composite parabolic or hyperbolic forms assemble in the same concave arcs, in a tapered pattern, or else in several parabolic forms, e.g. in the Rajasthan Desert, India (Fig. 15).

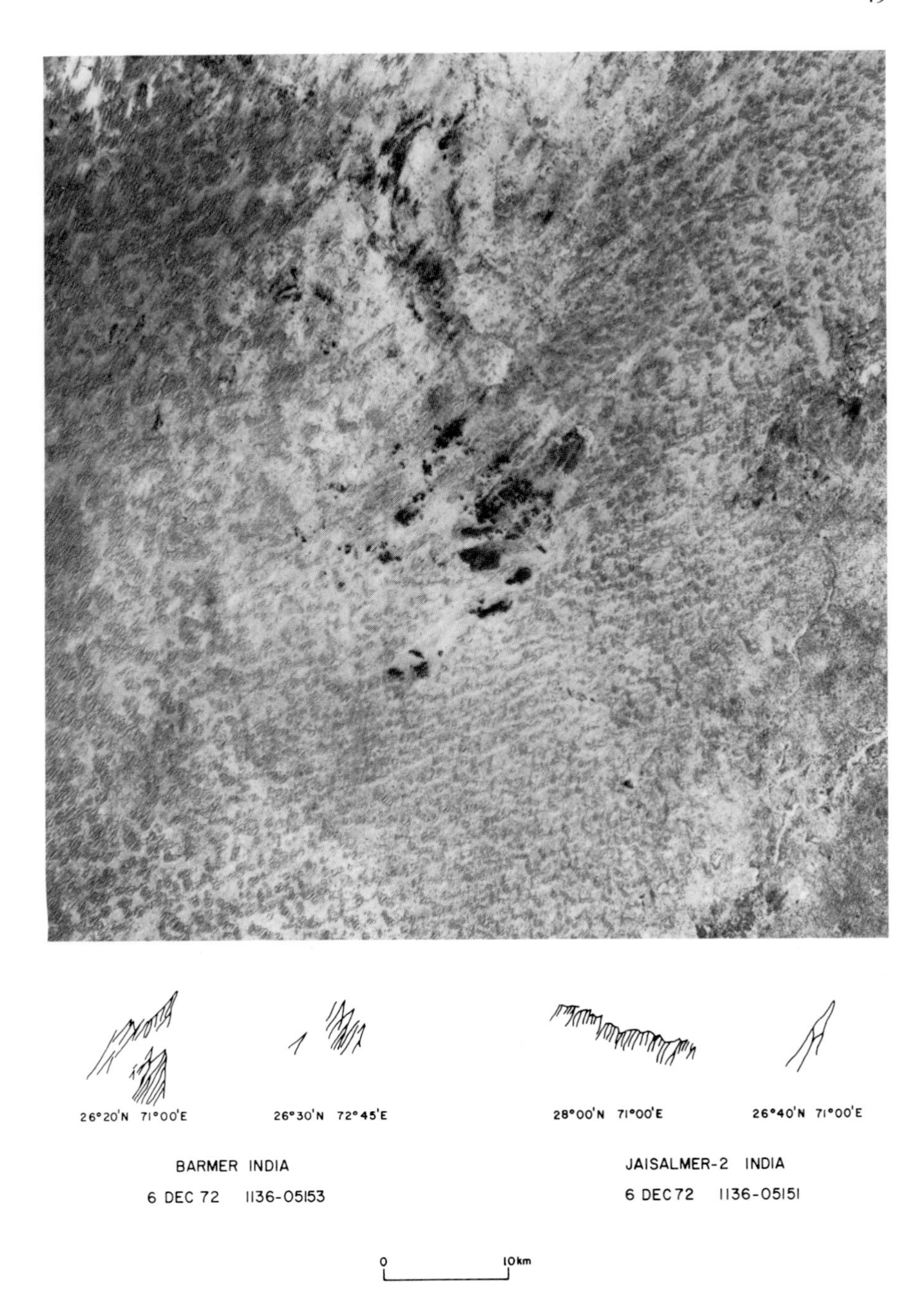

Fig. 15a. Landsat image of parabolic dunes in the Rajasthan Desert of India.
Fig. 15b. Illustration of compound parabolic dunes in western Rajasthan Desert, India.

Sandridges

Sandridges (residual dunes) are large, longitudinal ridges, separated by corridors formed by deflation. These are the most widely distributed of the terrestrial dunes; on Mars they are located in the circumpolar erg, only at one place 210° W–83° N, where they had been observed at the beginning of formation. The explanation for this scarcity resides both in the excess balance of sand feeding and in the young age of the erg. Many authors have described sandridges in the western Sahara [5, 6, 19, 28–30, 47, 52–58], in Libya and Egypt [40, 48–51], in Namibia [12, 56, 59], in Australia [60, 61, 64, 65], in Arizona [62], in Saudi Arabia [24, 63], and in the Erg of Fachi Bilma in Niger [25, 39].

Linear dunes and sandridges have often been classified in the same family, but this is an error because they are distnguished from one another by the dynamics of formation. In linear dunes, the migration of sand does not conform to the direction of the wind, but it arrives obliquely at the dune and is trapped, to be carried obliquely downwind. In 1950, Price [66, 67] realised that there was a confusion between linear dunes, those which were defined in 1941 by Bagnold, and sandridges, which were defined in Australia by Madigan. It was Price who emphasized the necessity to distinguish between the two.

Sandridges are the longest edifices fashioned by the wind on the surface of our planet. They are much wider than linear dunes; the latter have a scant base, which rarely exceeds 100 m in width. In a morphometric analysis of some ridges, Breed and Grow [12] showed that their width varies over tens of meters (from 43 m in the north of Arizona to 1480 m in the western Rub Al Khali). Measurements made on satellite images reveal that in the Chech and the Iguidi ergs of north Africa, complex ridges of more than 2 km in width occur. Although sandridges exist essentially in arid areas, they are also found in the Sahel, covered by steppe in Niger, in Upper Volta, in Mali and in Australia, where only the uppermost top layers have been reactivated.

Sandridges are always aligned in the direction of the dominant winds. In 1920, Chudeau [14] wrote 'the dunes of the Sahara form long ribbons closely parallel to the dominant winds'. This was also supported in 1930 by Aufrere [69]. Thus, they are the only forms that should be called 'longitudinal' dunes. This fundamental characteristic differentiates them from linear dunes, which are oblique to the wind and result from the 'sillage'.

In the Rombolara Hills of Australia's Simpson Desert these dunes have rather rectilinear paths. When they encounter a topographic obstacle, they swerve around it into a curved shape until they gradually resume their initial direction downwind.

The fact that these edifices are longitudinal to the dominant direction of the wind, is essential to the interpretation of aeolian phenomena on satellite images. These dunes are precise indicators of the direction of the dominant wind. However, such indicators must be thoroughly checked out, because on satellite images, sandridges and complex linear dunes of the feathered-bunch type may easily be confused with one another.

For example, Breed and Grow [12] do not sufficiently distinguish between longitudinal and linear dunes. This error stems from the exclusive use of satellite images, without adequate field checks. In the fields of sandridges, the anemometrical setting is complex, thus prohibiting the observation of these perfect alignments. For example, at the bendingof the Niger River in Mali, two series of ridges converge. North of Timbuktu (17° N), the sandridges are very close together, and trend NE-SW 67°. The same trend is observed at north Lake Niangay, 16° N and 3° E, where the ridges encounter another series of even larger (2.5 km wide) ridges from the ENE-WSW direction (88°). The situation reflects the encounter of the undeflected NE-SW trade winds blowing directly from the Sahara towards the Sahel, with the trade winds, which are reoriented in the Sahel in an E-W direction.

The gigantic sandridges of the Erg Chech trend in an NNE-SSW direction, nearly parallel to the dominant trade winds and the sand exportation. However, in the same area, there is another wind and the sand arrives from a direction that is oblique to the sandridges. This results in incoming barchans at the eastern side of the erg. In the interdune corridors, the ENE-WSW circulation of sand is expressed by striations that are transverse and oblique to these corridors, with obliquity from one corridor to another.

Observations in the Oubari Erg of Libya confirm that sandridges are immobile but not unalterable edifices. The center of the Oubari Erg consists of complex sandridges surmounted by feathered bunch types of linear dunes, and closed sinuous dunes in the form of a beehive or even of star dunes.

In this erg and north of Sebha, the Gaberaon Oasis owes its existence to the presence of a permanent lake in an interdunal corridor. The lake is just at the foot of a very steep (about 30°) southern slope of a sandridge, which dominates the lake without ever having an avalanche of sand from the summit into the lake. At first, it was impossible to determine how such a large edifice could thrive near the lake without ever losing vast quantities of its sand into it. Now we know that it was because of the direction of the sand exportation, which was parallel to the sandridge, rather than perpendicular or oblique to it.

While sandridges are edifices with the majority of the mass being immobile, important transformation of shapes can intervene. In the erg Chech, where the dominant wind parallels the sandridges, simultaneously, the local sand-carrying winds are transverse to the dunes and the corridors. It is these localized winds alone that explain the asymmetry of the sandridges at the erg Chech, which are near rectilinear on the eastern face where the sand arrives, and irregular on the western face where the sand escapes towards the following ridge. The maximal arrival of sand corresponds to this oblique direction, ENE-WSW, while the direction of the maximal removal is probably in the direction of the trade wind (NNE-SSW). Similarly, north of the Gilf Kebir plateau in the Western Desert of Egypt, sand is transported obliquely across the north-south trending corridors.

As noted above, the width of these edifices is essential to their definition, with the notion of erosional interdune corridors separating them also meriting attention. These dunes compose most of the terrestrial ergs. However, in addition to just their wavelengths, we believe that sandridges should also be studied as residual dunes. This is why the coefficient of depletion 'D' for an erg is such that:

$$D = \frac{W_i \text{ (width of an interdunal space)}}{W_r \text{ (width of a ridge)}}$$

If $D \leqq 1$, as in the erg Makteir (at the north of Richat, 21° 50′ N and 11° 35′ W), the erg of Azouad at the north of Timbuktu (17° N and 13° W) and the erg of Fachi Bilma in Niger, where the width of the sandridges has been measured as varying between 750–1500 m with an average of 1100, the function of D varies from:

$$3 > D > \frac{1}{7}.$$

All of these ergs are in the southern part of the Sahara. They are distinguished from those in the northern Sahara, where the coefficient D is largest.

Satelitte images of the erg Er Raoui and the erg Chech reveal a D coefficient growing in the direction of the trade wind.

at 29° N, $D = 4$
at 26° N, $D = 7$

This observation is also true for the erg Iguidi as well as all of the ergs of the northern Sahara, which seem to empty as if their principal source of feeding had disappeared.

Hanna [63] pointed out that the spacing between sandridges in the Kalahari and in Australia, which are five times smaller than those in the Sahara and Saudi Arabia, could be explained by the differences between the greater average speed and the height of the trade winds inversion. Satellite images confirmed this, but only the ergs of the northern Sahara have large interdunal corridors. Those ergs in the south are more densely occupied by dunes. Also, the ergs of Australia and the Kalahari are vegetated and the longitudinal dunes have already degraded. The sands from the dunes are without a doubt diffused in the corridors.

Having defined the sandridges as dunes of erosion leads to the question of how deflation scoops out the corridors. Aufrere [8] writing on this subject noted that a fluid in motion is not a dynamic homogeneous medium. When going from the top of the ridges towards the center of the corridors, the friction diminishes and the wind velocity increases. The maximal activity corresponds to the axes of the corridors, and the minimal activity corresponds to the axes of the chains. This explains the hollow regions of the corridors and the accumulations of sand on the ridges, but it does not explain the transverse rhythm impressed on the passage that one finds in a structure acquired from the wind.

Two principal hypotheses have been proposed to explain the genesis of the corridors: (a) the existence of helical air flows with a horizontal axis, turning in an inverse sense, returning the sand from the interdunal corridor to the dune, [60, 61, 63]; and (b) suggested by Aufrere, and later by Bourcart [11] and King [70], and supported in this paper as well, is that of deflation.

The proposed erosional hypothesis is similar to that of regressive erosion in water. It is supported by the interpretation of satellite images. In the same erg, the coefficient 'D' is often greater downwind than upwind. Longitudinal regressive exportation of sand can be combined with the helical effect from the air flow.

The sandridges, contrary to the transverse edifices, can only be shaped in the context of a negative budget of sand feeding. In the event of a positive budget of sand, an increase in the thickness of the sand sheets and their fashioning into akles or transverse chains preceeds the phase of a deficient budget, i.e., a growing aridity, creating less sand, and with increasingly powerful winds from which results a deficit of charge and an excess of energy. In Niger, where the erg of Fachi Bilma passes in the arid region of the Sahelian zone, the superposition of longitudinal grooves on transverse chains is quite common. It can also be seen north of Lake Chad (Fig. 6).

The types of sandridges can thus be considered according to: (a) their size (small in Arizona, middle sized in the Simpson desert of Australia, and large in the erg of Fachi Bilma and in the Check erg); (b) the degree of periodicity of the ridges; periodical ridges (as in the Simpson desert and in the erg of Fachi Bilma); and aperiodical ridges (as in Egypt in the Ghard Abu Muharik which is isolated, and in Upper Volta, where in the Sahel the ridges cross the entire northern part of the country, beginning even farther east in Niger); and (c) the presence (gassi) or absence (feidj) of sand in the corridors, (as in gassi Touil in the Erg Oriental and gassi in the Erg of Fachi Bilma, and feidj in the ergs Chech and Iguidi, Fig. 16).

Complex forms

Often sandridges become complex because of the superposition of linear, barchan, or star dunes. Linear dunes are superposed on sandridges in the erg of Fachi Bilma and the erg Check where the linear dunes are often sinuous and reclose upon themselves. Barchans or reversing dunes occur on sandridges, as in Arizona. In the southern central region of the state of Colorado, E. McKee pointed out sandridges that were badly separated from one another and covered by transverse reversing dunes, a result of the blowing of the wind for nine months

Fig. 16. Erg Iguidi, very spaced sandridges.

from the west and for several weeks from the east. Star dunes also occur on sandridges, as in the Oubari erg.

When linear dunes occur on sandridges, the two families of edifices are distinguished from one another by their direction, their profile and their grain size. The differences in direction result from the alignment of the sandridges in the direction of the main wind, with the linear dunes aligned obliquely to the same wind. Whenever they do not cover the entire sandridge, they are always found on upwind face. In the erg of Fachi Bilma they always extend from the NNW towards the SSE face. They sometimes increase in number such that they extend from the ridges to the corridors. The linear dunes are always composed of material finer than that found in sandridges.

The presence of linear dunes on sandridges can only be explained by a change in climate from a dry period with a rich influx of sand to a period during which the influx of sand decreases as the aeolian energy grows. This is the way in which sandridges are formed. The wind progressively produces a coarser pavement of sand by winnowing, totally fixing the sandridges, and protecting them against deflation. It is during the following phases that the linear dunes of fine sand appear on these fixed sandridges.

Conclusion

This classification of dunes distinguishes between the forms of pure accumulations of barchanic edifices, transverse chains, linear dunes and star dunes; erosional forms of parabolic dunes, and sandridges. These two classes of aeolian edifices appear to be responses to different sediment budgets. The excessive budget results from a wind saturated with sand and is responsible for the ergs of transverse chains. The deficient budget is due to a change in climate. For example, the case of an increasingly arid climate will result in the consumption of the sand which was weathered and removed from its original rock by previous fluvial activity. It is moved by the wind by increasing aeolian activity. This leads to a gradual emptying of the ergs, causing longitudinal dunes to become the dominant features.

Along the gradual evolution, as the ergs get older, their coefficient 'D' of depletion increases. This is true for active ergs which become emptier when the sources of sand are consumed. This corresponds to an evolution of the climate from wet or semi-humid to one of aridity, but is not the rule for fixed or semi-fixed ergs where dunes are degraded by the gradual spreading out of sand. This second evolution corresponds to the passage of a climate from aridity to semi-humidity, to finally a wet climate.

Acknowledgements

This paper was ably translated at the Smithsonian Institution, Washington D.C. by Ellen Lettvin through the courtesy of Dr. Farouk El-Baz, who also edited and shortened the original manuscript. Thanks also to Dr. Harold Dregne of ICASALS of Texas Tech University for showing me the parabolic dunes of Texas. Thanks also to the following staff members of the Center for Earth and Planetary Studies, Smithsonian Institution: Ms. Donna Slattery for typing the manuscripts, Lesley Manent and Rosemary Aiello, who helped to prepare the figures.

References

1. Aufrère L: Le problème géologique des dunes dans les déserts chauds du nord de l'Ancien Monde (Sahara, Arabie, Inde). Paris: C.R. Masson, Assoc. Fr. Av. Sci. 53e session, Le Havre:393–397, 1929a.
2. Aufrère L: Les dunes dans les deserts, Paris: La Météorologie, Série V:277–278, 1929b.
3. Aufrère L: Le cycle morphologique des dunes. Ann. Geogr. T.40 (226):362–385, 1931a.
4. Aufrère L: Le problème géologique des dunes dans les déserts chauds du nord de l'Ancien Monde, Paris: C.R. Somm. Soc. Géol. Fr., p 23, 1931b.
5. Aufrère L: Classification des dunes. Paris: A. Colin. C.R. du Congrès Int. de Géogr., Paris 1931, T II, Section 2, 1er fasc:699–711, 1933a.
6. Aufrère L: Les dunes continentales, leurs rapports avec le sous-sol, le passé géologique récent et le climat actuel. Int. Geogr. Congr. 13e Paris, 1931. C.R. 2(1):699–711, 1933b.
7. Aufrère L: Les dunes sahariennes et les vents alizés. B. Assoc. Fr. pour l'avancement Sci. 112:131–138, 1933c.
8. Aufrère L: Essai sur les dunes du Sahara algérien, Stockholm: Géogr. A, T. XVIII:480–550, 1935.
9. Bashin OJ: Dunenstudien. Gesellschaft für Erdkunde zu Berlin, Z. 6:442–430, Abstr. in Geogr. J. 23:588, 1903.
10. Beadnell HLL: Desert sand dunes. Cairo Sci. J. 3(34):171–172, 1909.
11. Bourcart J: L'action du vent à la surface de la terre. Paris: R. Géog. Phys. Géol. Dynam., T. I:26–54, 1928.
12. Breed C, Grow T: Morphology and distribution of dunes and sand seas observed by remote sensing. In: McKee ED (ed) A study of global sand seas, 1979, pp 253–305.
13. Capot-Rey R: La morphologie de l'erg Occidental. Alger: Tr. IRS, T. II:1–35, 69–106, 1943.
14. Chudeau R: L'étude sur les dunes sahariennes. A. de Géog. T. XXIX:334–351, 1920.
15. Clos-Arceduc A: Essai d'explication des formes dunaires sahariennes. Paris: Inst. Geogr. Nation, Etudes de photointerprétation 4:59, 1969.
16. Clos-Arceduc A: Disposition des structures éoliennes au voisinage d'un groupe de barkhanes à parcours limité, étude d'un groupe isolé de barkhanes au sud du Tibesti, évolution des barkhanes sur un parcours limité par deux aires où la déflation éolienne interdit la présence des dunes, 1971a.
17. Clos-Arceduc A: Typologie des dunes vives. Tr. Inst. Géogr. de Reims, 6:63–72, 1971b.
18. Cooke R, Warren A: Geomorphology in deserts, 1st ed. London: B.T. Batsford Ltd., 1973, 374 pp.
19. Duffour A (Lnt): Observations sur les dunes du Sahara Méridonal. Paris: A. de Géogr., T. XLV:276–285, 1936.

20. Fédorovic BA: Quelques considérations sur l'origine et le développement des reliefs sableux. Acad. des Sci. URSS, Série Géogr. et géophys. 6:885–910, (en russe, titre traduit, résumé en anglais), 1940.
21. Foureau F: Quelques considérations sur les dunes et les phénomènes éoliens. Mission Saharienne, Docum. Sci. T. II:213–237, 1903.
22. Foujita T: Note on sand dunes. US Army Corps of engineers/NASA, Earth resources survey from spacecraft, v. 2, appendix 1, 1967.
23. Grabau AW: A textbook of geology, part 1. Boston: Heath, 1920.
24. Holm DA: Sand dunes. In: Fairbridge RW (ed) The encyclopedia of geomorphology. New York: Reinhold, pp 973–979, 1968.
25. Mainguet M: Proposition pour une nouvelle classification des édifices sableux éoliens d'après les images de satellites Landsat 1, Gemini, NOAA3. Zeitschrift für Geomorphologie, Bd 20(3): 275–296, 1976b.
26. Matchinsky M: La formation des dunes dans les déserts. Nature 2341:1169–1175, 1953.
27. Petrov MP: Le relief des sables mobiles des déserts et des semi-déserts et les règles de sa formation. Tr. Inst. Geogr. AN SSR. n. 36, (en russe) 1940.
28. Rolland G: Sur les grandes dunes du Sahara. B. Soc. Géol. Fr. X(3):30–47, 1882.
29. Smith HTU: Classification of sand dunes. 19e Cong. Int. Alger 1952, Déserts actuels et anciens. Sec. VII fasc. 7, p 105, 1954a.
30. Smith HTU: Eolian sand on desert mountains. Geol. Soc. of Amer. B. 65:1036–1037, 1954b.
31. Solger F: Studien über norddeutsche Gulanddunen, Berlin: Forsch. Z. Dtsch Landes und Volkskunde, XIX:1–89, 1910.
32. Walker RG, Middleton GV: Facies models 9, eolian sands. Canada: Geoscience 4(4):182–190, 1977.
33. Fryberger SG: dune forms and wind regime. In: McKee ED (ed) A study of global sand seas. Washington DC: USGPO, 1979.
34. Monod TH: Majabat al-Koubra. Contribution à l'étude de l'Empty Quarter ouest-Saharien. Mém. Inst. Fr. Afr. Noire 52:406, 1958.
35. Howard AD, Morton JB, Gad-el-Hak M, Pierce DB: Simulation model of erosion and deposition on a barchan. Virginia: University of Virginia, 1977, p 77.
36. Högbom I: Ancient inland dunes of north and middle Europe. Geog. Ann. 5:113–243, 1923a.
37. Högbom I: Ancient inland dunes of north and middle Europe. Stockholm: Dissenbahen, 1923b, 244 pp.
38. Oldham RD: A note on the sand hills of Clifton near Karachi. M. Geol. Surv. India. T. XXXIV(1), 1903.
39. Mainguet M, Callot Y: L'erg de Fachi Bilma (Tchad-Niger), contribution à la connaissance de la dynamique des ergs et des dunes des zones arides chaudes. Paris: CNRS, Mémoires et Documents, v. 19, 1978, 184 pp.
40. Bagnold RA: The physics of blown sand and desert dunes. London: Methuen 1941, 265 pp (reéditions 1954, 1960, 1965).
41. Tsoar H: The dynamics of longitudinal dunes. Final Tech. Report, Eur. Res. Off. London: US Army, 1978.
42. Cornish V: Formations des dunes de sable. Bruxelles: Inst. Géog. Publ. Univ. Nouvelle, n. 2, 1900.
43. Harlé E: Dunes parallèles au vent sur la cote de Gascogne, C.R. sommaire. Soc. Géol.de France, 1912.
44. Harlé E, Harlé J: Mémoire sur les dunes de Gascogne avec observations sur la formation des dunes. Bull. Soc. Géogr. Min., Inst. Publique Imprimerie Nationale n. 146, p. 1920, 1916.
45. Jennings JN: On the orientation of parabolic or U-dunes. Geogr. J.G.B., T. CXXII (4):474–480, 1957.

46. McKee ED: Structures of dunes at White Sands National Monument, New Mexico (and a comparison with structures of dunes from other selected areas). Sediment 7(1):1–69, 1966.
47. Armstrong PW: Saharan sand and the origin of the longitudinal dune: a review. Geog. Rev. T. XL:464–465, 1956.
48. Bagnold RA: Journeys in the Libyan Desert. Geogr. J. 78:13–39, 524–535, 1931.
49. Bagnold RA: A further journey in the Libyan Desert. Geog. J.82:103–129, 403–404, 1933.
50. Bagnold RA: Libyan sands, travel in a dead world. London: Hodder and Stoughton, travel book club, in-8°, 1935a, 351 pp.
51. Bagnold RA: The movement of desert sand. Geogr. J. 85(4) 342–369, 1935b.
52. Beadnell HLL: The sand dunes of the Libyan Desert. London: Geog. J. (April), 1910.
53. Beadnell HLL: Libyan Desert dunes. Georg. J., T. LXXXIV (4):337–340, 1934.
54. Besler H: Der Namib-Erg und die Sud-afrikanische Randstufe der Erde neu entdekt. Mainz: L. Beckel et S. Schneider, B 1, 25, 1975a.
55. Besler H: Messungen sur Mobilität von Dunensanden an Nordrander Dunen-Namib (Sudwest-afrika) Würzburg: Wurzburger Geogr. Arb., Heft 43:135–147, 1975b.
56. Capot-Rey R: L'Edeyen de Mourzouk, Inst. de Rech. Sahariennes, tr., T.IV:61–109, 1947a.
57. Capot-Rey R: L'exploration de l'erg Oriental (decembre 1944–janvier 1945) Tr. Inst. de Rech. Sahariennes, T. IV:181–188, 1947b.
58. Clos-Arceduc A: Essai d'explication des formes dunaires Sahariennes, Paris: Inst. Géogr. Nat., études de photointerprétation 4, 1969, 66 pp.
59. Dufour A (Lnt) Observations sur les dunes du Sahara méridional, B. de l'Assoc. de Géo. Fr., 88:84–88, 1935.
60. Folk RL: Longitudinal dunes of the northwestern edge of the Simpson Desert, Northern territory, Australia, I: Geomorph. and grain size relationships, Sediment 16:5–54, 1971a.
61. Folk RL: Genesis of longitudinal and ghourd dunes elucidated by rolling upon grease. B. Geol. Soc. Amer. 82:3461–3468, 1971b.
62. Hack JT: Dunes of the western Navajo country. Geogr. Rev. 31:240–263, 1941.
63. Hanna SR: Formation of longitudinal sand dunes by large helical eddies in the atmosphere. T. of Applied Met. 8(6):874–883, 1969.
64. Mabbutt JA: A striped land surface in western Australia. Trans. Inst. Brit. Geogr. 29:101–114, 1961.
65. Mabbutt JA, Sullivan ME: The formation of longitudinal dunes: evidence from the Simpson Desert. Australian Geogr. 10:483–487, 1968.
66. Mainguet M, Callot Y: Air photo study of typology and interrelations between the texture and structure in the erg of Fachi Bilma, Sahara. Geomorphic processes in arid environments. Proc. of Jerusalem – Eilat symp. Berlin: Z. fur Geomorph. A Geomorph Supp, Bd 20(1):60–69, 1974.
67. Price WA: Saharan sand dunes and the origin of the longitudinal dunes; a review. Geogr. Rev. 40(3):462–465,1950.
68. Aufrère L: L'orientation des dunes continentales. Int. Geogr. Cong. 12e, Cambridge 1928, Report of Proceedings: 220–231, 1930.
69. McKee ED (ed): A study of global sand seas. Geol. Surv. Prof. Pap. 1052, 1979, 429 pp.

Author's address:
Laboratoire de Géographie Physique Zonale
Université de Reims
Reims, 51100 France

3. Space observation of Saharan aeolian dynamics

Monique Mainguet

Abstract

With images provided by satellites, a new analytical tool for observing the landscape of a region is now at our disposal. In the field of aeolian dynamics, satellite data have provided a new means for the study of sand seas and the interpretation of the direction of sand transport. These newly perceived relationships between the Sahara and surrounding climatic zones and the dynamics between the Sahara and its semi-arid margins are discussed in this chapter.

Introduction

A flight over the Sahara by airplane or the study of Landsat images reveal a paradoxical desert, because the sand of the Sahara is actually somewhat scarce. Striking is the abundance of rocky outcrops and the density of the fluvial valleys. A close examination reveals that the rocky outcrops are exaggerated by the presence of desert varnish. Sandstones are rocks which are easily veiled by this varnish and they are also richly represented in the Sahara.

The prevalence of old fluvial valleys is also a result of desert mechanisms. Although these valleys are not active, except occasionally, they are visible merely because they are floored by aeolian sand deposits which are trapped in the valley in the midst of their transport by the wind.

However, despite these two aspects, the Sahara is a desert which can best be defined by its aeolian parameters. Ever since the major climatic change that made the Sahara what it is presently, wind has been the dominant geomorphological factor. The climatic change was in the Miocene; according to Jaeger [1]. Louvet and Magnier [2], and Sarnthein [3]. For Maley [4] as well, the Sahara was not completely formed as a desert until the end of the Miocene. Geomorphological arguments revealed by field studies [5] also substantiate the idea that the Sahara is an ancient desert. At the periphery of Tibesti, there exists a *Kalutian* system (aeolian corridors and crests measuring hundreds of kilometers

El-Baz, F. (ed.), Deserts and arid lands. ISBN: 90-247-2850-9.

in length for each kilometer of width). In these corridors, there are at least three different lacustrine phases. The depth of aeolian sculpture in these hard, Paleozoic sandstones could only have been accomplished over tens of millions of years. In fact, it appears that the same aeolian systems have recurred throughout history after each humid phase. The direction of the barchanic convoys that are now being displaced are in conformity with ancient systems. However, in this article emphasis will be placed on present aeolian dynamics as well as rejuvenated ancient systems.

The paradox of the Sahara is that it is a desert relatively poor in sand whereas in the Sahel, its meridial margin, the sand sheets attain incredible thickness; 60 m were measured north of Maradi in Niger. This astonishing situation is a direct result of the aeolian currents of sand transport in the Sahara, which trend towards the Sahel and the Atlantic Ocean. This tendency, if permanent will result in a desert devoid of sand, but this is not the case. It is thus important to introduce the temporal parameter and the paleoclimatic alternation – of dry and of more humid phases, during which weathering made coats of alterites, part of which are winnowed by the wind and built into ergs during drier phases. Aeolian production of sand by corrasion of rocks constitutes a minor sand source for the ergs.

Satellites images are a major analytic tool in studies of this type, but alone they are insufficient. Aufrere [6] first suspected aeolian currents based on his field work. Prior to the advent of satellite imagery, aerial photographs were the only remote sensing tools. These pictures also showed evidence of transsaharan currents [5]. Since the advent of satellite imagery, studies of the east of the Sahara have been expanded to cover the entire desert. The results from aerial photographs and satellite images have since been substantiated by field observations in Algeria, Morocco, Chad, Niger, Upper Volta, and in the laboratory by sedimentological studies.

Dynamics of climatic zones

The dynamic relations between the Sahara and the other climatic zones is of special significance. Climatologists have indicated that it is no longer possible to regard climatic regions or zones as being independent of one another: there is a dependence on latitude, longitude and altitude. Similarly, there is an aeolian interdependence between the Sahara and its semi-arid borders as well as with the rest of the planet. The Sahara is the world's largest source of dust. The currents exporting aerosols have been followed on meteorological satellite imagery: [7–14].

The movements of dust that they clarified are (a) towards the northern, temperate (Europe including Iceland and Ireland), peri-glacial (Greenland) zones [15, 16, 17]; (b) towards the southernmost sectors (Sahel, the Sudan zone and the Guineas zone – we believe that the Gulf of Guinea's evergreen forest, its luxuriance and resistance, must in part be due to the deposition of Saharan dust, since aerolian particles have been found in the soils of the forest zone of the Gulf of Guinea by both geologists and pedologists [10, 18, 19]; (c) towards the west, supported by numerous American scientists, who first recognized, and later measured the aeolian exportations of the Sahara towards and across the Atlantic to the Bermudas, Cuba, Brazil, and even beyond [9, 11, 20–26]; and (d) towards the east; the middle east and Arabia even receive dust from the Sahara [27, 28].

North-south transsaharan relationships

Also important in the relationship between the Sahara and other climatic zones in the north-south transsaharan wind regime. Although it is known that deflation and saltation leave characteristic traces on the ground, these have only recently been identified [5, 19]. By simultaneously using aerial photographs and satellite images of Landsat and Meteosat followed by subsequent field checks, it is now possible to prove the existence of transsaharan crossings. In these crossings, the sand is submitted to a general displacement from north to south (Fig. 1). Using geomorphological criteria, a map of the paths of transsaharan and transsahelian transport has now been compiled by analyzing the images furnished by Meteosat from May 20, 1978 to January 25, 1979. This map (Fig. 1) shows how the Sahara and the Sahel are influenced by the NE-SW currents of sand transport, conforming to the trade winds in the absence of topographical obstacles. The currents bend at the latitude of the Tropic of Cancer and blow across the Sahel, changing from an ENE-WSW direction to E-W.

The map of the paths of sand transport conform completely to a map analyzing the July winds of the Sahara established by Lamb [29] (Fig. 2). It is also the first recognized as conforming to the great paths of migration followed by migratory birds and locust infestations across the Sahara. This is a good example of the intuition of animals for using the wind to carry them, facilitating their long trips. This same continental scale of transport can also be deduced from the works of Brookfield [30] in Australia (Fig. 3), of Holm [31] in Saudi Arabia (Fig. 4), and of studies in South Africa.

The functional heterogeneity of the paths of sand transport can be seen distinctly on Landsat images. These currents consist of alternating sectors: those where sand transport is dominant and those where deposition dominates. It is in the sectors dominated by deposition that sand seas thrive. Thus, chains of ergs are materialized, as in the following examples:

The first chain (Fig. 5) is the westernmost of the Sahara, uniting the Grand Erg Occidental to the erg Er Raoui, as well as the erg Iguidi. The latter erg covers a

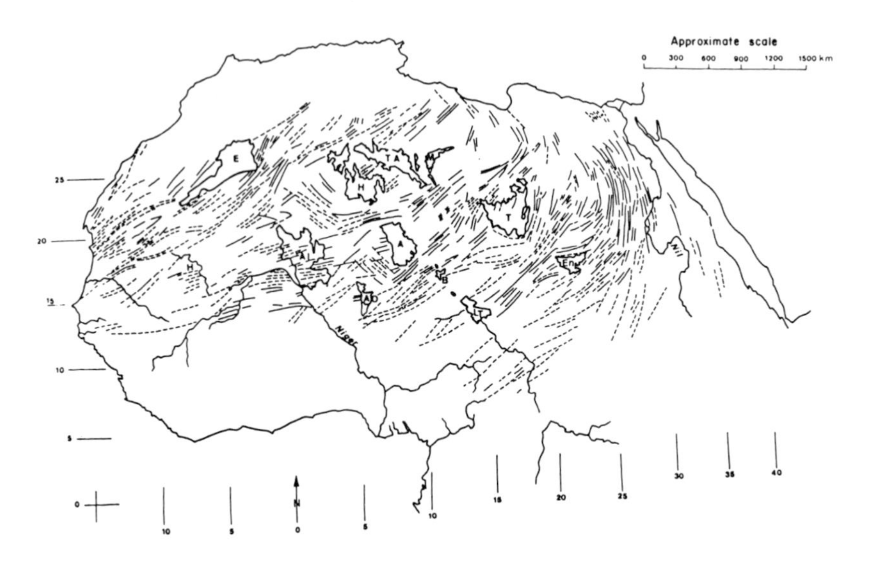

Fig. 1. Transsaharan and transsahelian aeolian currents based on data recorded from Meteosat between 5/30/78 and 1/25/79; A: Air; A.D: Ader Doutchi; A.I.: Adrar des Ifoghas; B: Bilma; E: Eglab; En: Ennedi; H: Hoggar; H: Hodh; L.T.: Lake Tchad; M: Messak; T.A.: Tassili N'Ajjer; T: Tibesti. ——— = Lines of sand deposition or corrasion (discerned by strong reflectance on Meteosat images); ----- = Lines of deflation (weak reflectance on Meteosat images).

large area of deflation and transport which connects with the dunes of Fort-Gouraud, which are in turn connected to the scarcely vegetated Azefal erg by a surface where the sand transport has created striations typical of corrasion, before reaching the Atlantic near 20° N.

The second chain (Fig. 5) again is situated to the west, emanating from the Great Western Erg (Grand Erg Oriental), connected to the erg Er Raoui, which is followed by the erg Chech which also receives a substantial amount of sand from the Grand Erg Oriental. In the course of its transport across the Tademait, it impresses marks of corrasion in the ground. Starting south of the erg Chech two branches develop: *first*, figures of deflation and of corrasion across the Hank imply a relationship with the linear dunes of Makteir, and also with the Akchar erg which is covered with a thin steppe, before ending at the ocean near 19° N; *second*, the erg Check (Fig. 5) is connected with the erg Ouarane which is comprised of transverse dunes. Then it connects with the vegetated ergs of Amoukrous and Aoukar that also connect with the vegetated erg of Trarza by means of linear dunes moving across a ridge-corridor south of Atar.

It is not presently known how the general transport is arranged in the center of the Sahara, except for the existence of circular currents at the periphery of Hoggar from the east, the south and north of this massif. To the north is the

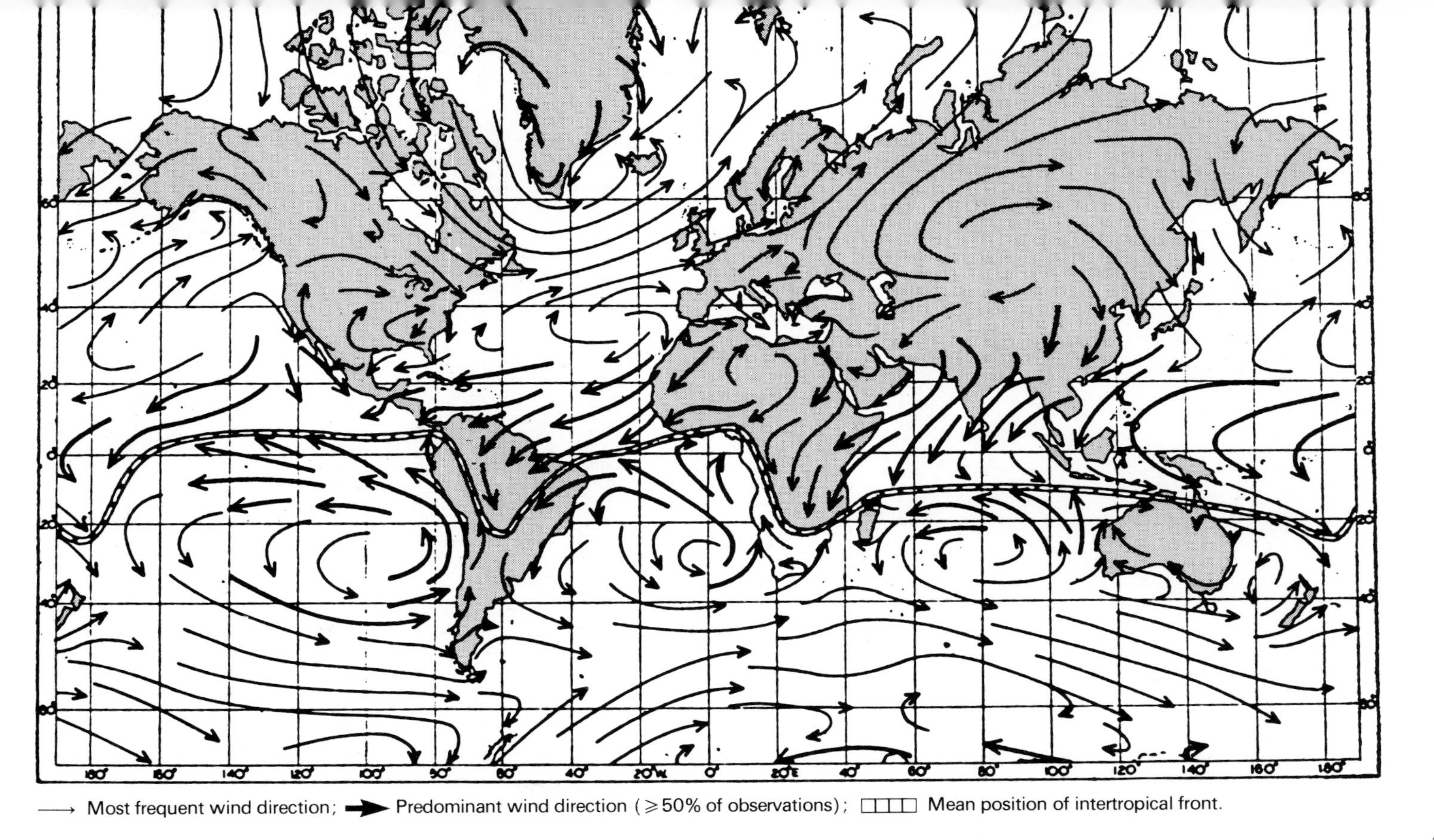

⟶ Most frequent wind direction; ➡ Predominant wind direction (≥ 50% of observations); ▭▭▭▭ Mean position of intertropical front.

Fig. 2. Prevailing surface winds in January (in period between 1900–1950) [29].

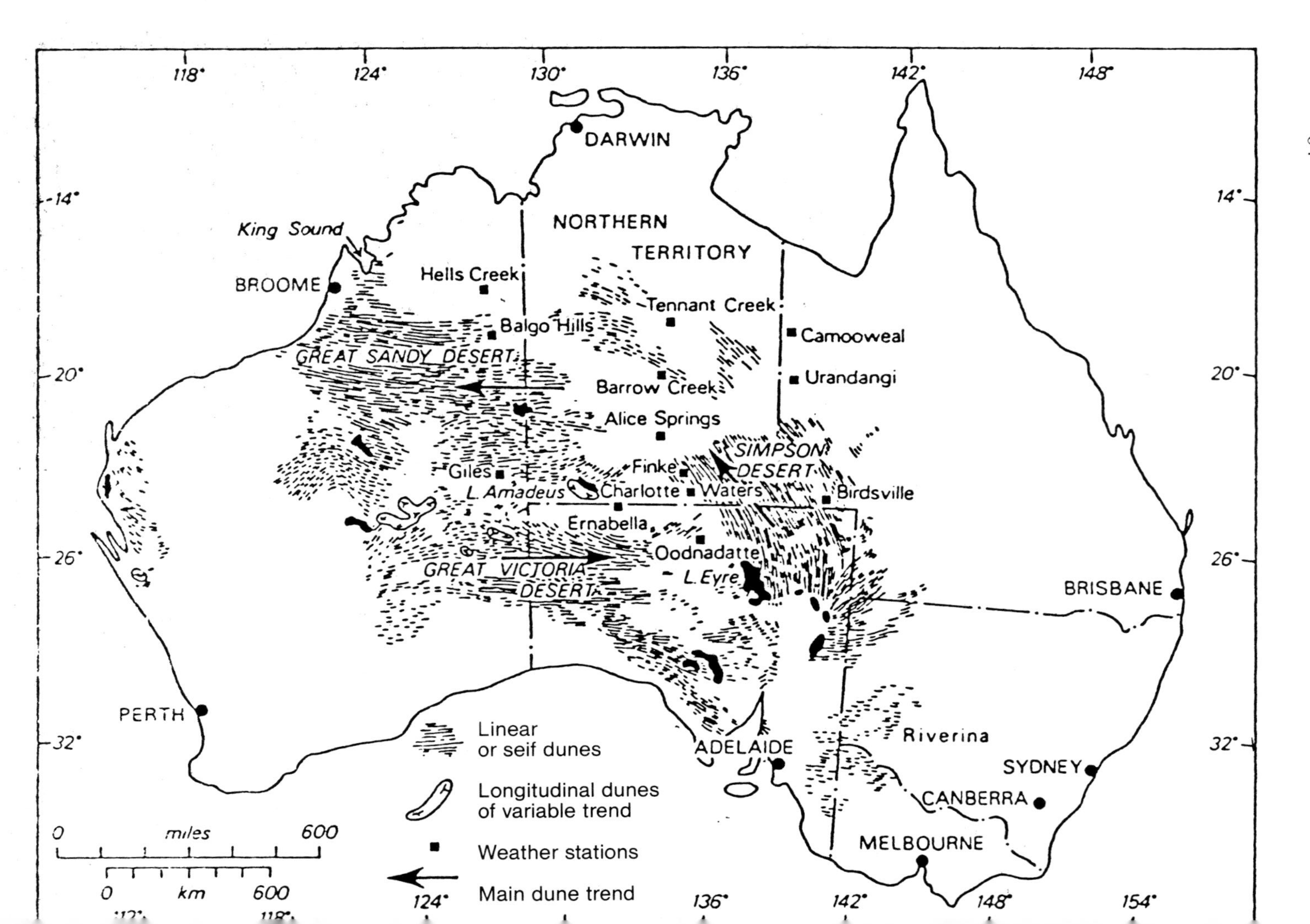
118°
124°
130°
136°
142°
148°
14°
20°
26°
32°
DARWIN
NORTHERN
TERRITORY
King Sound
BROOME
Hells Creek
Balgo Hills
Tennant Creek
Camooweal
Urandangi
Barrow Creek
Alice Springs
GREAT SANDY DESERT
SIMPSON DESERT
Giles
Finke
L. Amadeus
Charlotte Waters
Birdsville
Ernabella
Oodnadatte
L. Eyre
GREAT VICTORIA DESERT
BRISBANE
PERTH
Riverina
ADELAIDE
SYDNEY
CANBERRA
MELBOURNE
Linear
or seif dunes
Longitudinal dunes
of variable trend
Weather stations
Main dune trend
0
miles
600
0
km
600
124°
136°
142°
148°
154°

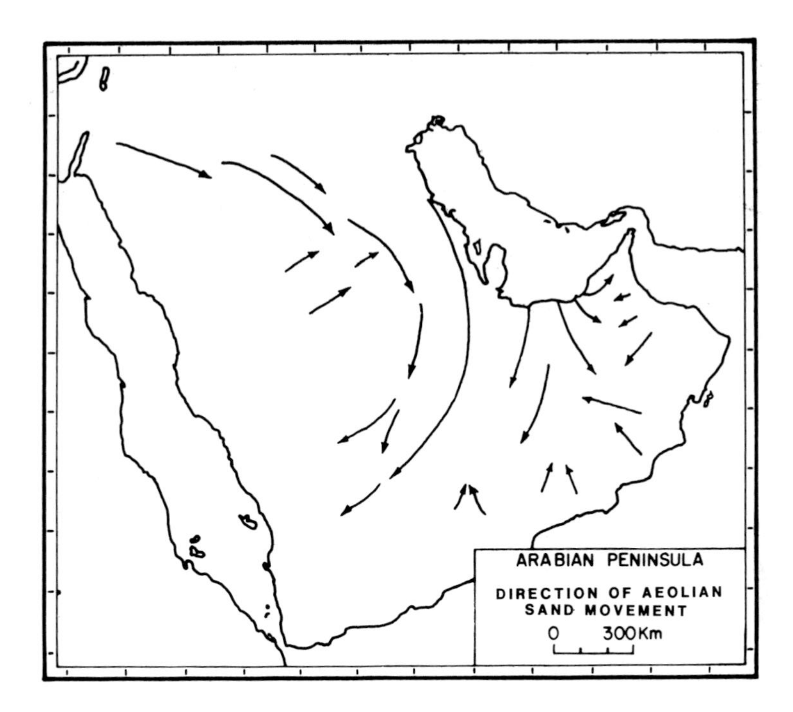

Fig. 4. Main wind currents in the Arabian Peninsula [31].

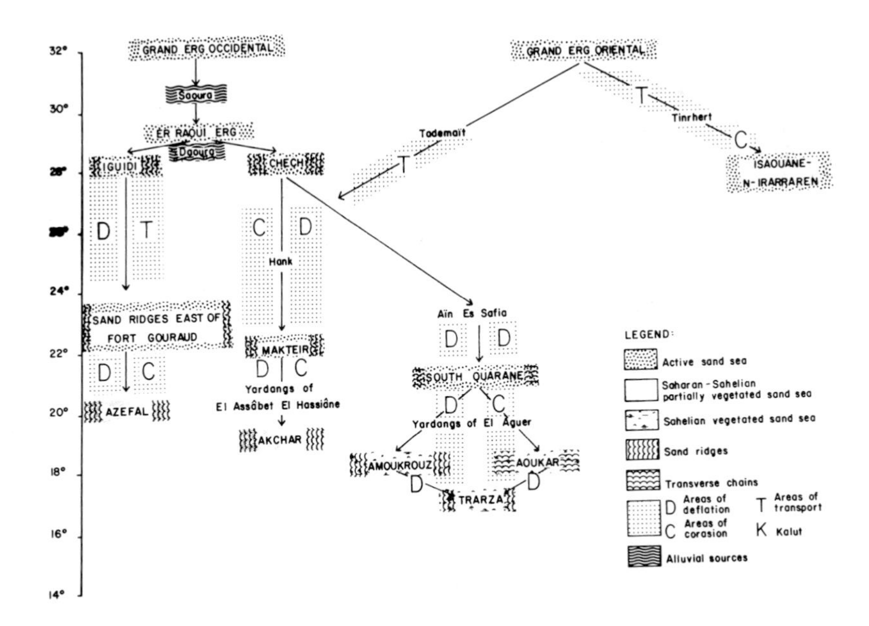

Fig. 5. Erg Chains in the Western Sahara.

circulation across the Tademait and the Tinrhert which are believed to be chiefly SW-NE [32].

The analysis of images provided by the NOAA-4 satellite (Fig. 6) has revealed circular movement verifying data from Landsat images. From the latter, the deviation towards the northeast of the aerolian paths at the southwestern edge of Hoggar, and even a SW-NE transport (proceeding in the opposite direction of the trade winds) has been confirmed at the central southwest of Hoggar and between the west of Tassilis and the plateau of the Tademait. Nonetheless, analysis is not yet advanced enough to affirm that these diverse fragments pertain to a functioning unity proceeding counter-clockwise around Hoggar.

In the eastern part of the Sahara (Fig. 7) the aeolian circulation near the ground is influenced by the Tibesti massif [5, 33]. The Landsat satellite images have provided use with the following associations: The first course of aeolian sand displacement from the north to the south of the Sahara is delineated by the Calansho and the Rebiana ergs, the Great erg of Bilma, the vegetated Hausa erg and the great sandridges of Niger and of Upper Volta (Fig. 7); The second course goes from the Great Sand Sea of Egypt to the Egypto-Sudanian goz. From there, the sand is either displaced to the ergs situated at the east of the massif of Ennedi, or else it is displaced to Djourab, and the Kanem. After Lake Chad, the current rejoins the preceding chain and crosses the Hausa erg (Fig. 7). General remarks were made arising from the observation of the sand displacements along these currents.

The above observations were deduced from satellite images, and were compared with data acquired from the examination of aerial photographs. Sedimentological studies of aeolian sand in the laboratory, and field survey of terrain led to the following general remarks regarding sand transport across the terrain:

First, the transport of sand is discontinuous in time and space. The quantity carried is determined by the amount of available movable material and by the speed of the wind, which is highly influenced by the topography. Sand seas are found at sectors of lesser aeolian energy, and in areas where slopes occur oriented against the wind direction.

Second, the alternation of sectors dominated by deflation with those dominated by deposition. Deflation occurs when the active load of sand in the wind is less than the capacity of transport, and when sand is made available from corrasion on the surfaces of the rocks. The most characteristic forms of corrasion are produced in the sandstone [5, 34, 35]. We will ultimately see that in an erg, when the arrival of sand is less than the loss, there is a negative overall effect and the prevailing forms of the dunes are longitudinal sand ridges.

Third, locally in the Sahara, the aeolian circulation, which is globally NE-SW, is affected by the continental trade winds from one side and by the topographical environment at the other. The ergs Chech and Iguidi are tangible

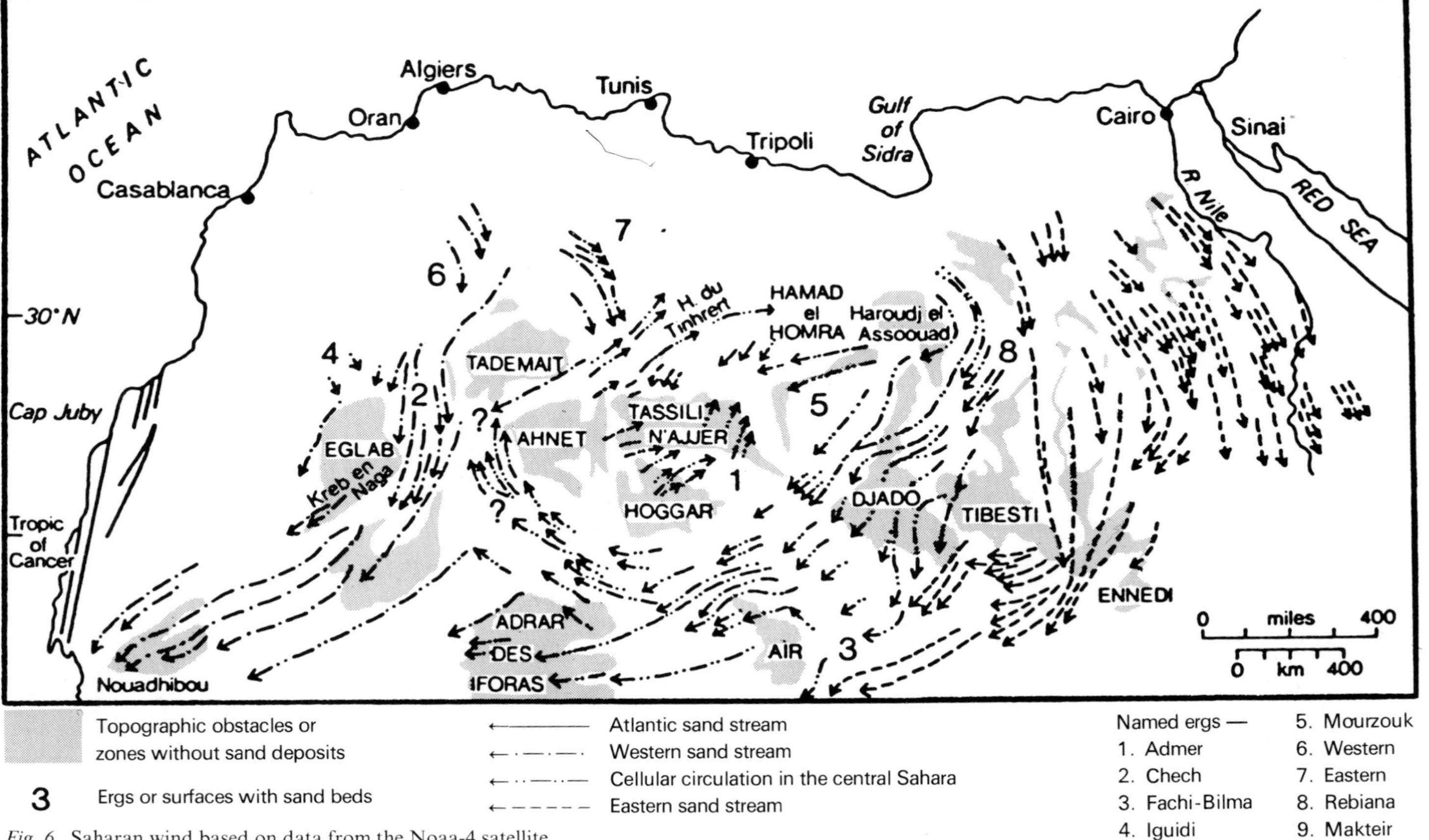

Fig. 6. Saharan wind based on data from the Noaa-4 satellite.

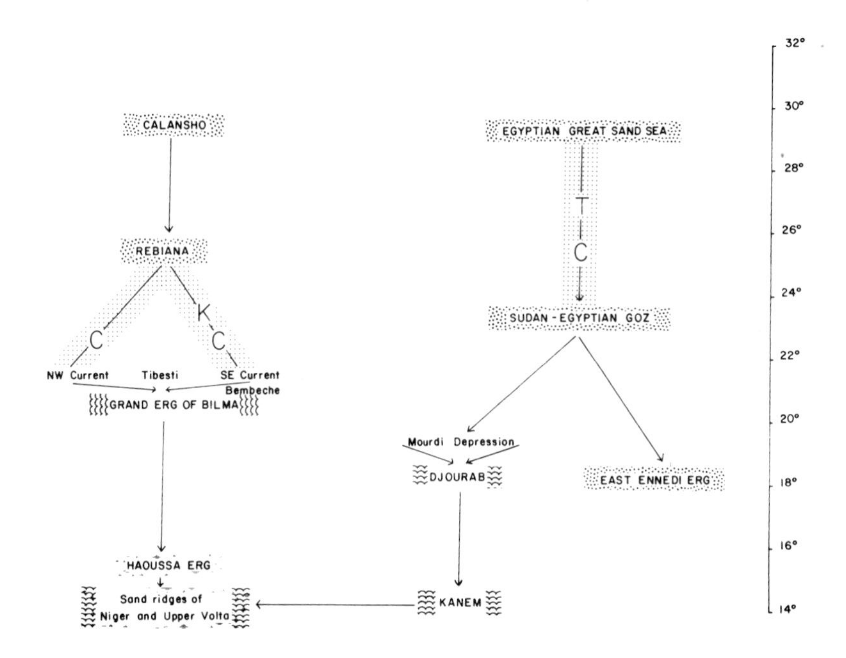

Fig. 7. Erg Chains in the Eastern Sahara.

evidence of how the Eglab is responsible for the subdivision of the aeolian circulation of the sand into two branches (Fig. 8). It is the same around Tibesti and around Hoggaar, where a turning circulation develops.

Fourth, the perception of another phenomenon on the scale of the entire Sahara is induced by the observations of a general translation of the majority of sand towards the south. The Sahara evidently furnishes particles to the Sahel. This explains the extent and the thickness of the sand cover in the Sahelian zone. To the north of the Sahara, in the moroccan oases of Draa, Zis and Rhesis, the same movement is originated by a wind called 'Saheli' issuing from the Sahara bringing the sand back from the southwest towards the northeast. A certain percentage of sand arriving in the oases, which still remains to be determined, is probably of Saharan origin. The wind can have a Saharan origin and the transported sand can originate from any place along the path of the wind, with the prevalent wind depositing material along its trajectory step by step.

Ergs of the Sahara

An erg is a vast dune assemblage functioning as an area of sand arrival, deposition, and stockage. However, the importation of sand to it must be, or have been in excess of its exportation.

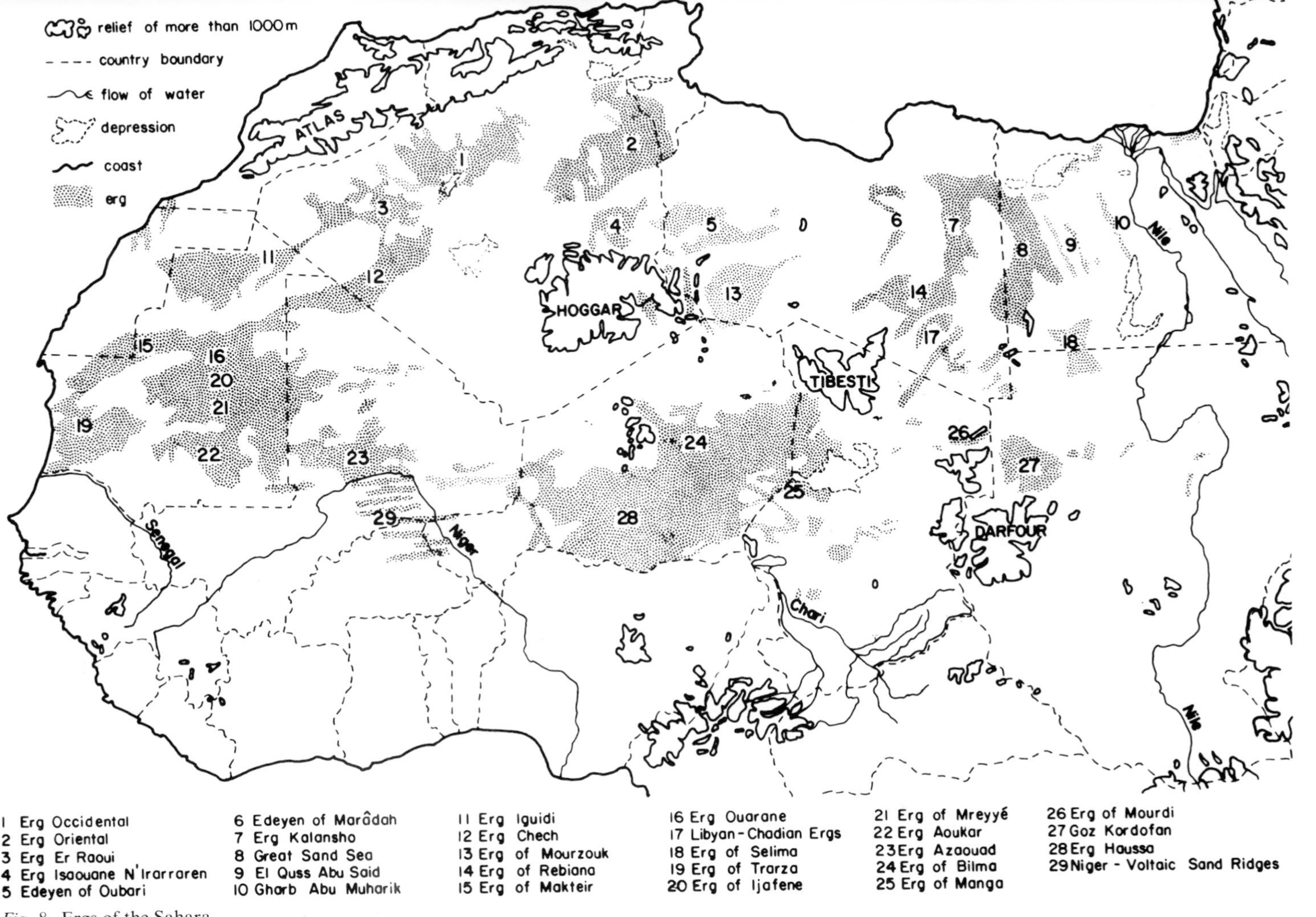

Fig. 8. Ergs of the Sahara.

In order to understand an erg, the three following items must be clarified: (a) the way in which sand progresses towards the erg. (It is probable that the present modes of progression are different from those of the past, when fluvial feeding was of greater importance. However, in this study only the aeolian means of transport will be considered); (b) the way in which the sand is concentrated and localized in the erg; and (c) the specific nature of the dunes of which an erg is composed.

Previous authors, in order to explain the materialization and growth of ergs, concluded that the feeding of sand is entirely autochthonous. They considered an erg to be like a field of dunes resulting from a simple reworking of sand, carried in a shallow basin by hydric transport. This denied the capacity for long distance transport of the wind [6, 36–42]. Lieutenant Dufour, when observing the Sahara by camel, was the only person since 1936 to notice the existence of great movements of sand in the meridial part of the Sahara. Aerial photographs and satellite images have indicated that the feeding of the sand may be allochthonous and/or autochthonous. Mineralogical and morphoscopical analyses confirm both possible origins for the sand of ergs.

With a predominantly monodirectional aeolian regime, a planar topography and an influx of sand equal to the possibilities of export by the wind [42], barchans, aerodynamic vehicles of sand transport are the result; in more complicated aeolian regimes, on the rough topographic surfaces or on sand sheets modeled into waves, linear dunes are formed (linear depositional dunes, oblique to the wind) along which the sand, temporarily trapped by the edifice, is transported as if along a rail.

Dunes the size of barchans and linear dunes (seifs) are often too small to be discernible on satellite images, and thus this dynamic scale escapes observers. When these dunes reach a sufficient size, it is possible to perceive them on satellite images. This observation is favored by the strong reflectance of these edifices. Examples of this are in Egypt, from the south of the Qattara depression to the Tropic of Cancer, where due to the exceedingly planar topography and the regular flow of winds from a dominant direction an intense transport of sand occurs in two forms: massive organized transport in barchans, and grain by grain saltation across immense aeolian plains.

The barchans are visible due to their giant size: they are more than 500 m from one arm to another. The breaking up of such large barchanic waves in the Kharga depression destroys villages and plantations [43].

Saltation and rolling can be deduced by the corrasion traces they leave behind. In such areas aeolian action is marked by surfaces sculptured in yardangs, especially when an escarpment interposes itself on the passage of sand, as in the case of the escarpment of Kharga where the sculpting of yardangs is presently active.

An example of the dynamics of an erg in the north of Niger in the Saharo-Sahelian zone is that of Ténéré and the Hausa ergs (already described by Urvoy [44–48] and by Grove [49]. They both consist of sandridges and transverse chains. Noticeable on the Landsat images are some seifs of reactivation which are much more reflective than the total surface of the sand seas.

Localization of ergs

An erg usually forms due to the presence of an opposing slope along the path of migration of saltating sand. This is why, in basins, ergs do not begin to grow in the bottom most part, but rather, after having crossed the bottom of the depression, the material accumulates at the foot of the opposite slope. This phenomenon is also true on a local scale: as soon as a branch of one of these great currents encounters a small obstacle, a deposit is formed, which acts as a springboard, raising particles upslope.

On Mars, to the south of the peripheral erg surrounding the polar ice cap are volcanic craters in which small ergs exist, forced against the upwind slope by the wind, without totally occupying the craters. At the south of the Sahara, the erg of Djourab (Chad) does not occupy the Bodélé depression, the most depressed area in the shallow basin of Chad, rather, it is situated in the southern most region, where the substrate raises itself slightly. Similarly, the Great Sand Sea of Egypt adjoining the Qattara depression to the west, is situated on a great elevated surface.

The second most significant aspect of the localization of ergs is the direction and velocity of the wind. This is the case for ergs at the southern border of the Sahara, where deposition results in the slowing down of the trade wind. An obstacle or the presence of vegetation causes the ergs to thicken towards the Saharo-Sahelian zone. Also, the encounter of two currents of opposite direction which nullify each other result in ergs. The best example of this is the erg of Fachi Bilma, a sandy area of 800 km situated southwest of the foot of the Tibesti massif, under the trade winds. As soon as the NE-SW trade wind encounters Tibesti which is more than 400 km in diameter, it divides itself into two branches, each carrying its own load of sand. These two branches contour the massif, one to the northwest, the other to the southeast, before rejoining each other leeward of the massif. Here they nullify one another and deposit their loads of sand in the shadow of the obstacle [33]. This effect is observable up to nearly 500 km leeward of the Tibesti massif.

Models of erg formation

Models that pertain to the formation of ergs of the Sahara basically explain the transverse and linear dunes. The synoptic view provided by satellite images enables the classification of dunes into three fundamental families: (a) transverse dunes and chains; (b) star dunes, medusal or dome dunes with or without arms; and (c) sandridges and linear dunes.

Transverse dunes can exist in any form from isolated crescents to transverse chains in the course of their evolution, spaced progressively closer together until an 'akle' structure is formed. The Sahara does not offer a good example of this type of erg. In contrast, in the west of the Ala Shan desert in China (40–40° 30′ N, 100–101° E), isolated crescentic dunes proceed into transverse chains, growing in the upwind direction. The erg of Rub al Khali, in Saudi Arabia (21°–22° N, 54°–55° E) is of the same family, composed of megabarchans and gigantic barchan chains. Anemometrical measurements and the construction of wind roses support the two following propositions: as the dune's sand is deposited in an aeolian regime with one dominant direction, the ergs are comprised primarily of transverse dunes; if they are formed in a multidirectional aeolian regime, they consist of ghourd type pyramidal dunes [42]. The best examples of this type of erg is in the North Polar region of the planet Mars (Fig. 9), where isolated barchanic edifices at the peripheral limit of the sand seas proceed in the direction of the wind in this same area, coalescing barchanic edifices forming chains, which pass progressively into 'aklés'.

Star dunes are more or less pyramidal or convex, they either possess many arms and a medusal form, or a simple dome with or without well formed arms. The Erg Oriental contains some of the most beautiful star dunes on Earth. Starting at the south of a line between El Oued and Touggourt, in an area crossed by moving sand, star dunes trend towards 33° N, arranged in chains from the western part of the erg. Later these chains trend N-S and at the south of Hassi Messaoud, they get separated by large 'gassis' or corridors, one of which, the Touil gassi, serves as a principal path for N-S trade routes in eastern Algeria. To the east of this gigantic area of star dune chains, the Erg Oriental is formed by star dunes, which change from the dome shape to a pyramidal shape. All of the transitional steps between dome dunes with arms which are not well developed and pyramidal, long armed star dunes are represented in the same area.

Sandridges range from simple to complex types, by the superposition of transverse barchanic crescents or active linear dunes oblique to the wind. The Check and Iguidi ergs, composed primarily of ridges, provide examples of this. Transverse dunes are the response to a wind carrying excess sand, where deposition of sand exceeds the exportation. Sandridges, however, are the result of a completely opposite situation, where erosional forms become prevalent. This happens most frequently downwind of sand seas, where the interdune corridors

Fig. 9. Transverse chains in the great north polar erg of Mars (81° N, 141° W).

grow larger by a regressive erosion, propagating in a manner similar to that of a fluvial environment. This explains why the southwest terminus of the erg of Fachi Bilma, downwind of the massif of Termit at the Niger, displays wide interdune corridors, much wider than those situated at the heart of the erg.

The two northernmost Saharan ergs, Erg Oriental and Erg Occidental, are still in a phase where the input of sand exceeds the output. This results in very dense dunal assemblages, expecially in the northern part of the erg.

In the south central Sahara, the ergs are comprised in part of longitudinal undulations that are located in the flow of the circulation of the trade wind, upwind of the transverse chains specific to the Sahel.

Thus, all of the ergs of the Sahara are at different stages of evolution. Those comprised mostly of longitudinal undulations or sandridges, with interdunal corridors are situated at a central latitude of the desert. In the Iguidi and Chech ergs, these corridors attain maximal size: ten times wider than those found between the dune chains which run on a 'reg' towards the ergs of Mauritania and the Maqteir. These ergs are also formed of longitudinal ridges. Another branch is directed towards Ouarane, which is formed by transverse chains.

Active ergs in the south of the Sahara are replaced by vegetated ergs in the Sahel, coinciding with the latitude at which the action of the active winds approaches its limit.

In the central and south Saharan ergs, erosion is presently the dominant phenomenon creating sandridges, following the direction of the wind. This takes place in the vegetated ergs of the Sahel, which are mainly composed of transverse dunes and chains. The latter of these corresponds to a decrease in the velocity of the wind combined with the excess load of sand derived from the erosion of the upwind regions of the ergs at the south of the Sahara. It is also due to an increase in the rugosity of the substrate caused by the growing steppe of the Sahel. In the line of the wind, as it unloads the sand, its active force increases once again, and deflation becomes possible. All of these phases are illustrated in the erg of Fachi Bilma, where chains of star dunes oriented in the direction of the wind preceed in linear, regularly undulating sandridges. When the ridges arrive in the Sahel, they evolve into different types of transverse edifices as evidenced in the Hausa erg. They are followed by nigeriovoltaic sandridges separated by corridors of deflation, from Ader Doutchi to the plain of Gondo at the foot of the cliff of Bandiagara. These sandridges are aperiodical, because the sandy cover is less dense, discontinuous and the topographic irregularities (small fragments of plateaux, levelling of rocks, large N-S valleys) are numerous.

The Viking orbiter images of Mars show that in the giant circumpolar erg (5×10^5 km^2), there are very few linear dunes; this erg is apparently dominated by barchanic edifices, isolated or organized into barchanoid and transverse chains. This can be explained either by an excessive influx of sand due to seasonal variations, or because the polar cap released rocks which had previously submitted to powerful mechanisms of reduction, which the polar winds of the summer had carried as debris. These eastern winds, under the effect of the coriolis force, bring sand to the erg between 80° and 83° of north latitude and 100–240° east longitude on a polar projection map.

On the side where the development of the erg is maximal, the glacial cap does not advance beyond 80°; and on the other side, where the erg is less vast, it descends up to 76°. The notion of a budget for measuring sand growth, is expressed by the function $B = I/E$, where I is the influx of sand into an erg, and E is the export of sand out of the erg. This relationship clarified the equilibrium between sand import and export which is fundamental to explaining the forms of the dunes of which sand seas are composed, as well as the motives within the ergs.

If $B > 1$, the resultant dune will essentially be transverse, under a regime dominated by unidirectional winds; if $B < 1$, the dunes will consist of longitudinal dunes, and they will represent erosional forms. In other similar ergs, the relationship of B can change over the course of geological time or else it can change as a result of climatic changes, in which case the erg changes its structure.

Conclusion

The view from space of the Sahara reveals the general aeolian unity of that desert for the first time. However, in the ergs, the regions of sand accumulations are relatively small as compared with the regions of sand displacement between the dunes. Satellite images from the north and central areas of the Sahara show that the ergs, except in the extreme northern fringe, are areas where the dunes occupy less space than the interdunal regions. The sand seas of these north and central parts are becoming depleted of their sand.

The center of the Sahara has particularly few ergs. One probable reason for this is the presence of mountains. Here, the continental trade winds encounter the massifs of Tibesti and Hoggar, dividing them into several branches. In each of these regions the wind accelerates and the aeolian energy increases; this is an unfavorable condition for sand deposition and for the growth of ergs. On the contrary, in the south of the Sahara and the north of the Sahelian zone, there are ergs formed predominantly of densely populated transverse sandy edifices. This situation is due to the large amount of sand coming from the north and deposited leeward of the wind path, marking the end of its transport.

The notion of sand budgets clarifies this general trend which is manifested in all deserts. In the Sahara, the principal mechanism of shaping the landscape is aeolian action. All of these parameters combine to form a single dynamic, and functioning unity, directed by the trade winds and the topography. In the Sahara, the terrain is strewn with major obstacles which dictate the positions of the ergs by accelerating the winds along the lateral flanks, and decelerating the wind in the downwind direction. This same arrangement of a single, functional unity of aeolian phenomena also exists in other continental deserts with smoother topography, as in Australia [30], Saudi Arabia [31] and in South Africa.

References

1. Jeager JJ: Les rongeurs du Miocène moyen et supérieur du Maghreb, Thèse Sci. Montpelier, Fasc. 1, 1975, 1964 pp.

2. Louvet P, Magnier P: Confirmation de la dérive du continent africain au Tertiaire par la paléobotanique. Toulouse: Science 96ᵉ cong. nation. soc. savantes 5:177–189, 1971.
3. Sarnthein M: Sand Deserts during glacial maximum and climatic optimum. Nature 272(5648): 43–46, 1978.
4. Maley J: Les changements climatiques de la fin du Tertiaire en Afrique: leur conséquence sur l'apparition du Sahara et de sa végétation. In: Williams NAJ, Faure W (eds) The Sahara and the Nile. Quaternary environments and prehistoric occupation in northern Africa. Rotterdam: Balkema, 1980, pp 63–86.
5. Mainguet M: Le modelé des grès. Problèmes généraux, 'Etudes de photointerprétation'. Paris: IGN 9, thèse doctorat d'état, 657 pp.
6. Aufrère L: Morphologie dunaire et météorologie saharienne. Ass. Géogr. Fr. (56):34–48, 1932.
7. Bertrand J, Baudet, Drochon A: Importance des aérosols naturels en Afrique de l'ouest. J. Rech. Atmos: 846–860, 1975.
8. Drochon A: Commentaires sur un phénomène de brume sèche observée par satillite ASECNA (37), DEM, 1970.
9. Jaenicke R: Atmospheric aerosols and global climate. J. Aerosol Sci. II: 577–588, 1980.
10. Kalu AE: The African dust plume: its characteristics and propogation across west Africa in winter. In: Morales C (ed) Saharan Dust, 1979, pp 95–118.
11. Mainguet M, Canon L, Chemin M-C: Le Sahara: géomorphologie et paléogéomorphologie éoliennes. In: Williams NAJ, Faure H (eds) The Sahara and the Nile. Quaternary environments and prehistoric occupation in northern Africa. Rotterdam: Balkéma, 1980, pp 2–35.
12. Morales C (ed): Saharan dust: mobilization, transport, deposition, rep. workshop, Sweden: Gothenburg, 1977.
13. Morales C (ed): Saharan dust. (Chichester) New York: Wiley, 1979, 297 pp.
14. Noyalet A: Utilisation des images de Météosat, genèse et évolution d'une tempête de sable sur l'ouest africain. Direction de la météorologie. Lannion, 1978, 8 pp.
15. Bucher A: Retombées des poussières d'origine africain (inst. et obs. de globe) Clermont-Ferrand, réunions de travail sur les caracteristiques physico-chimiques et le transport des pousssières d'origine africaine. Rés. de commun. inédit.
16. Nalivkine TV: Orages, tempêtes et trombes de sable. La Science, Leningrad, 487 pp (in Russian).
17. Sticher H, Bach R, Brugger H, Vokt U: Atmospheric dust in four soils on limestone dolomite and serpentine (Swiss Jura and Swiss Alps) Catena 2:11–22, 1975.
18. Goudie A: Dust storms and their geomorphological implications. J. of Arid Environments Im: 291–310, 1978.
19. Mainguet M, Cossus L, Chapelle A-M: Utilisation des images satellites pour préciser les trajectoires éoliennes au sol, au Sahara et sur les marges sahéliennes. Interprétation des documents de Météosat (5/26/78–2/9/79) B. Soc. Fr. Photogramm. 78:1–15, 1980.
20. Carlson TM, Prospero JM: The large scale of movement of Saharan air outbreaks over the northern equatorial Atlantic. J. Met. 11:283–297, 1972.
21. Delany AC, Delany CA, Parkis DW, Griffin JJ, Goldberg ED, Reimann BEF: Airborne dust collected at Barbados, Geochim. and Cosmochim. Acta 31:885–909, 1967.
22. Idso SB: Dust storms. Scientific American 235(4):108–111, 113–114, 1976.
23. Jaenicke R: Monitoring and critical review of the estimated source strength of mineral dust from the Sahara (SIES). Sweden: Gothenburg, Workshop on Saharan dust transport, 1977, 8 pp.
24. Judson S: Erosion of the land. Amer. Sci. 56(356), 1968.
25. Prospero JM, Carlson TN: Vertical and areal distribution of Saharan dust over the western equatorial north Atlantic ocean. J. Geophys. Res. 77(27):5255–5265, 1972.
26. Schultz L, Jaenicke R: Aeolian dust from the Saharan desert. In: Palaeoecology of Africa and the surrounding islands. Rotterdam: Balkema, 12:97–98, 1979.
27. Vinogradov BV, Grigor'ev AA, Lipatov VB: Directions du transport régional des particules

solides dans l'atmosphère des régions arides d'après les donnés obtenues par télédetection. Izv. Akad. Nauk. SSR sur Geogr. 4:78–84, 1978.
28. Yaalon DH, Ganor E: East mediterranean trajectories of dust carrying storms from the Sahara and Sinai (reprint- workshop on Saharan dust) Sweden: Gothenburg, 1977.
29. Lamb HH: Climate: Present past and future, I: Fundamentals and climate now. London: Methuen and Co. Ltd., 1972, 613 pp.
30. Brookfield M: Dune trends and wind regime in central Australia. Z. Geomorph. Suppl. 10: 121–153, 1970.
31. Holm DA: Sand dunes. In: Fairbridge RW (ed), The encyclopedia of geomorphology. New York: Reinhold, 1968, pp 973–979.
32. Mainguet M, Canon L: L'action du vent dans les paysages arides d'Afrique nord équatoriale, vue par les satellites, colloque GDTA: utilisation des satellites en télédétection, Saint Mandé 1977:307–1978.
33. Mainguet M, Callot Y: L'erg de Fachi Bilma (Tchad-Niger), contribution à la connaissance de la dynamique des ergs et des dunes dans des zones arides chaudes. Paris: CNRS, M. et Docum. 19:184 pp, 1978.
34. McCauley JF, Grolier MJ, Breed CS: Yardangs of Peru and other desert regions. US Dept. of Int. Geol. Surv. 1977. 177 pp.
35. Canon L, Galichet: Etude de télédétection de la dynamique éolienne de transport, d'accumulation et de corrasion entre l'erg de Mourzouk et l'erg d'Admer à travers les Tassilis N'Ajjer (M. de Maitrise). Reims, France: Université de Reims, (inédit) 1975.
36. Aufrère L: Le cycle morphologique des dunes. A. de Géogr. T. 40, 226:362–385, 1931a.
37. Aufrère L: Le problème géologique des dunes dans les déserts chauds du nord de l'ancien Monde. Paris: CR Somm. Soc. Geol. Fr., 1931b.
38. Aufrère L: Essai sur les dunes du Sahara algérien. Stockholm: Geogr. A. T. XVII, 1935.
39. Capot-Rey R: La morphologie de l'erg Occidental. Tr. IRS, Alger T. II:1–35, 69–106, 1943.
40. Capot-Rey R: Remarques sur les ergs du Sahara, Paris: A. Géogr:413, 2–19, 1970.
41. Chudeau R: L'étude sur les dunes Sahariennes, A. de Géogr. T. XXIX:334–351, 1920.
42. Fryberger SG: Dune forms and wind regime. In: McKee ED, (ed) A study of global sand seas, 1979, Washington DC: USGPO, 1979.
43. El-Baz F: The western desert of Egypt, its promises and potentials. In: Bishay A, McGinnies W (eds) Advances in desert and arid land technology and development I: 67, 1979.
44. Urvoy Y: Les formes dunaires à l'ouest de Tchad. A. de Géogr. 506–515, 1933a.
45. Urvoy Y: Modelé dunaire entre Zinder et le Tchad, Ass. de Géogr. Fr. B. 69, 10e année, 79–82, 1933b.
46. Urvoy Y: Terrasses et changements de climats Quaternaires à l'est du Niger. A. de Géogr. 44:254–263, 1935.
47. Urvoy Y: Structure et modèle du Soudan Français. A. de Géogr. 45:19–49, 1936.
48. Urvoy Y: Les bassins du Niger, étude de géographie physique et paléogéographie. Inst. Fr. d'Afr. noire, mémoire 4, 1942, 139 pp.
49. Grove AT: The ancient erg of Hausa and similar formations on the south side of the Sahara. Geog. J. 24:528–533, 1958.

Author's address:
Laboratoire de Géographie Physique Zonale
Université de Reims
Reims, 51100 France

4. Landforms of the Australian deserts

J.A. Mabbutt

Introduction

The major relief features of arid Australia are explicable in terms of large-scale geologic structure and tectonic history, whereas the patterns and status of the main depositional landforms are more attributable to past and present-day climates and to the dependant geomorphic processes.

The Australian deserts belong mainly to the *shield-and-platform* morphostructural type [1] and have many geomorphic features in common with the deserts of other Gondwana plate fragments, such as southern Africa. Crustal stability and a long sub-aerial history are reflected in a predominance of plainlands or plateaux on exposed crystalline basement rocks, and of tablelands or structural plains on the flat-lying cover rocks of unfolded intracratonic basins. The uplands, isolated and restricted within the broad plains, are generally related to ancient orogenic belts or to uplifted cratonic basin structures of ancient, resistant rocks. Although scenically spectacular, they are of modest relief, with the highest point, Mt. Zeil in the Macdonnell Ranges west of Alice Springs, barely attaining 1500 m.

The combination of tectonic stability and a long sub-aerial history with changing climates is also reflected in a widespread survival of Cenozoic palaeoforms. For example, the interior plateaux of Western Australia and of central Australia bear lateritic cappings and weathering profiles which result from planation under wetter conditions in the early Cenozoic, and the associated or derived soils are deeply leached and acid, with kaolinite-illite clay complexes. In the Interior Lowlands tributary to Lake Eyre are found mid-Cenozoic silcrete duricrusts. Still younger are the calcreted valley tracts and terraces, and tablelands of lacustrine limestones, manifestations of a disorganization of drainage linked with drier conditions in the later Cenozoic. Finally, the extensive dunefields indicate still more arid conditions in the Pleistocene although they are now mainly stabilized by vegetation.

Climatically the Australian deserts are only moderately arid, with no station recording below 100 mm for an average annual rainfall. The modest uplands do

El-Baz, F. (ed.), Deserts and arid lands. ISBN: 90-247-2850-9.

not induce orographic rainfall in geomorphologically significant amounts, and all local rivers are ephemeral. Even where streamfloods enter the arid zone from bordering subhumid uplands, as in north-central Queensland, rivers such as Coopers Creek do not have regular seasonal floods. The limited upland relief and extensive plainlands have led to an extensive disorganization of surface drainage in the arid interior, hence very little runoff from arid Australia now reaches the sea. On the other hand, the whole of arid Australia is topographically open to the ingress of tropical rainfalls, with the result that most stations have recorded well above their average annual rainfall in a 24 h period, when significant local fluvial action can occur.

Consistent with this moderate aridity, aeolian sand surfaces are generally vegetated, with the main exceptions being the deep sands on exposed dune crests and areas of significant active sand transport by wind, as for example on the northern (lee) shores of playas in the Lake Eyre-Lake Frome region. Aeolian landforms generally remain confined within the lower parts of the drainage basins within which the sands were initially deposited by fluvial action: there are no instances of dune systems extending across major drainage divides, as for example, in the hyper-arid Sahara.

Under the prevailing temperature regimes frost action is insignificant, and there are no relict periglacial or glacial landforms.

Physiographic divisions and types

The Australian arid zone occupies most of two of the major physiographic divisions of the continent, namely the Western Plateau and the Interior Lowlands [2]. The former consists of the cratonic basement of granitic rocks, outcropping in the continental shield areas, and sedimentary cover rocks of the intracratonic basins, ranging from moderately folded to flat-lying. The shield exposures and the little-disturbed platform cover together form the low plateaux, mainly between 300 and 500 m above sealevel, which characterize the western half of Australia. Folded and uplifted rocks of the older basins form the main uplands, including the parallel ridges of the Flinders and Macdonnell Ranges on steeply-dipping strata, and the bold plateaux of the Kimberleys and Arnhem Land on uplifted but little-deformed basin structures in the north.

The low gradients and locally insignificant watersheds of the Western Plateau have led to an advanced disorganization of former exoreic drainage systems. In part, the lower sectors of these systems have disappeared beneath a cover of aeolian sand; for example the Great Sandy Desert in the northwest extends across an old outlet of a river system which rises in the central Australian ranges. Elsewhere, broad shallow valleys across the shield plains are marked by belts of calcreted alluvium, and in Western Australia are occupied by lines of discrete playas blocked off by alluvium or aeolian sand – the so-called 'river lakes'. Many of the lower-lying plains of this division are occupied by dunefields or sand-plains.

This is the oldest part of the Australian land mass and it shared its earliest sub-aerial history with other contiguous plates of Gondwanaland. Lateritic relicts on the shield plains of western and central Australia, as old as early Cenozoic, indicate remarkably little geomorphic change over a long period of geologic time. The erosional landscapes were broadly fashioned under more humid climates in the past, perhaps under a savanna morphogenic regime.

The Interior Lowlands correspond to a tectonic depression between the craton of the Western Plateau and the uplifted orogen of the Eastern Uplands, and they mark the extent of an epeiric Mesozoic marine transgression. The floor is comprised predominantly of little-deformed Cretaceous shales and weak lithic sandstones, but the lower parts are extensively masked by alluvium and by fields of longitudinal sand dunes.

The chief physiographic province of the Interior Lowlands is the Lake Eyre Lowlands, with a centripetal interior drainage directed to Lake Eyre (– 16 m), the lowest point of the Australian continent. The survival of organized drainage in this province owes much to the basin disposition of weak and relatively impermeable bedrock, and also to the higher rainfall at the upland rim of the arid zone located to the northeast. Nevertheless, the summer streamfloods, diverted into alluvial basins in the lower river courses, and increasingly obstructed by extending dunes, reach Lake Eyre only every 5–10 years, whilst the lake can be expected to fill only 3 or 4 times in a century.

The main relief builder in the Lake Eyre Lowlands is a capping of siliceous duricrust above weathered profiles in the Cretaceous and Cenozoic rocks, preserving mesas and stony tablelands, particularly in the west-central part of the Lowlands. The disposition of these uplands has been influenced by domal uplifts, this being one of the few areas of significant neotectonic relief in Australia.

Further south, the northwestern part of the Murray Basin falls into the arid zone, comprising the alluvial plains of the Murray and Darling Rivers as well as their ancestral systems, and increasingly to the west, the Mallee dune-fields. Between these two domains is a zone of fluctuating and contested fluvial and aeolian dominance, with playas and source-bordering dunes at the terminals of the part-disorganized river systems.

Although gross morphostructural divisions within arid Australia usefully illustrate controls at the continental scale, they are too broad to provide a framework for a review of individual desert landforms. For this purpose a finer subdivision is required, and this account relies on the recognition of recurring *desert physiographic types*, each of which are made up of characteristic land-form groupings. These types constitute a sphere of dominant action of a

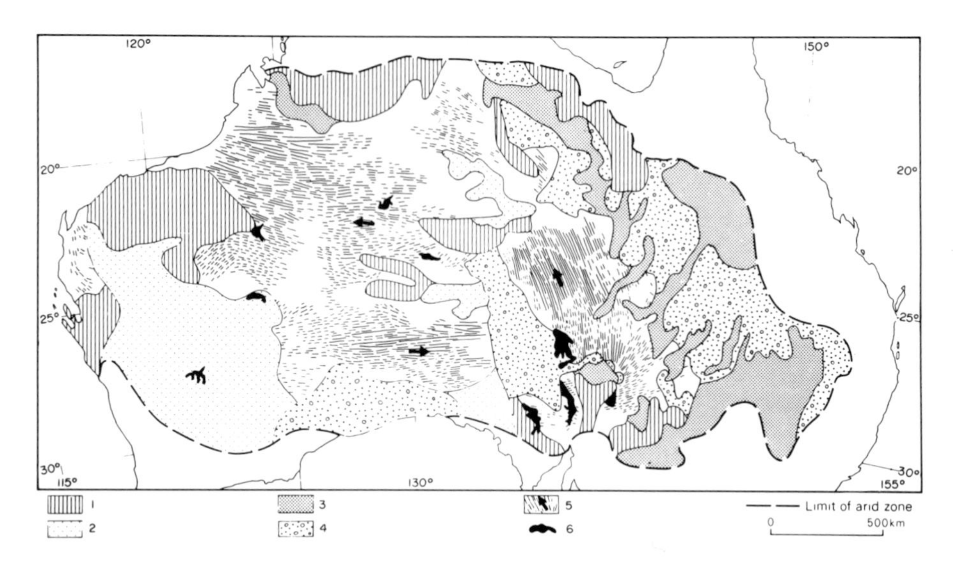

Fig. 1. Physiographic desert types in Australia: 1: upland and piedmont desert; 2: shield desert; 3: riverine desert and clay plains; 4: stony desert; 5: sand desert, showing trend and extension of longitudinal dunes; 6: desert lakes.

particular complex of related geomorphic processes. In the Australian deserts these patterns are sufficiently extensive to allow the mapping of generalized desert physiographic types at a broad scale (Fig. 1).

Upland and Piedmont desert

Only the larger uplands are visible in Figure 1, but their features are shared by many smaller relief islands. Despite their modest altitudes, the Australian desert uplands are significant as sources of local runoff, for only here are slopes steep enough, with sufficient convergent relief to provide the necessary gathering grounds for streamfloods. Functionally included with the uplands are the piedmont zones, which receive ephemeral streamfloods and their deposits from the adjacent uplands.

The main desert uplands are seen to be on the Western Plateau, and there are a variety of forms reflecting structural differences. Bold plateau relief occurs in the Kimberley Plateaux and Hamersley Ranges, where strata have been uplifted with little deformation, and strike ranges with ridge-and vale topography where they have been steeply tilted, for example in the Macdonnell and Flinders Ranges. Quartzites and sandstones are the main relief builders, but limestones are also prominent locally, as in the eastern Macdonnell Ranges. More complex upland forms occur where basement rocks have been folded or uplifted. Schists, foliated gneisses and slates tend to give a close grained relief of sharp ridges, and more rounded, broader summits are produced from granitic rocks. An example of this is in the Mount Isa Highlands of northwestern Queensland. Overall, such relief lacks the orientation of uplands developed on sedimentary rocks.

On most desert uplands, rocky outcrops with coarse blocky mantles alternate to a varying degree with smoother hillslopes of finer debris (Fig. 2). The former manifest features typical of selective and superficial desert weathering, and regimes which are generally weathering-limited in the sense the steeper slopes tend to be swept clear of weathering products as fast as they are yielded. Sandstone and quartzite faces generally exhibit reddish-brown crusts, characteristic of Australian desert uplands, behind which granular disintegration proceeds, as evidenced in honeycomb weathering and cavernous forms behind case-hardened surfaces.

Many of the ridges and plateaux in the Australian desert uplands exhibit a rough summit bevelling or subdued rounding. This is best exemplified in the Macdonnell Ranges of central Australia (Fig. 3). These have been interpreted as evidence of prior partial planation, and they are commonly associated with a subsequently incised, transverse drainage. Several of these upland bevels appear to be of at least Mesozoic Age, since they are continuous with the sub-Cretaceous unconformity beneath the adjacent Interior Lowlands. In general, there is an absence of contemporary deep weathering profiles in the hard sandstones which characteristically preserve there upland bevels.

Piedmont forms in these desert uplands typically include flights of gravel-capped terraces, as in the Macdonnell Ranges [3] or mantled pediments, as in the strike values of the Flinders Ranges [4]. The highest and oldest of these may exhibit a degree of ferricreting. Successions of piedmont gravel terraces have generally been attributed to a combination of tectonism and climatic fluctuations. Locally, higher benches capped with siliceous or ferruginous duricrust preserve the piedmont junctions of still older and higher plainland surfaces.

Shield desert

The occurrence of this desert physiographic type is determined by extensive outcrops of granitic basement rocks within the ancient structural division of the Western Plateau, and it includes some of the most ancient of Australian land surfaces. A long subaerial history combined with greater tectonic stability explains two important features of a shield desert, namely the great extent of plainlands, and the wealth of features inherited from a geologic past extending back to Mesozoic times, and from the moister environments that then existed.

The largest area of shield desert occupies the southern part of inland Western Australia. By early Cenozoic times this area had been reduced to a gently

Fig. 2. Upland and piedmont desert: uplands in arid Australia are of modest relief and generally stand isolated in open desert plains, as in the western Macdonnell Ranges. The subdued crestal forms, alternation of rocky and smooth slope sectors and abrupt slope break to the stony piedmont are characteristic. Photo by CSIRO.

Fig. 3. Upland and piedmont desert: gorge of the Finke River through the western Macdonnell Ranges, central Australia. The transverse drainage is probably inherited from an old planation surface here preserved as bevels across the sandstone strike ridges.

undulating land surface with broad shallow valleys, tributary to the Indian and Southern Oceans. The anomalous position of the continental drainage divide, near the south coast of Western Australia, suggests that this surface was fashioned when the Australian continental plate was continuous with that of Antarctica to the south. An extensive lateritic crust formed above weathered kaolinitic profiles in the uniform granite and gneiss defines this 'Old Plateau' surface. The Old Plateau has subsequently been shallowly incised and dissected, probably as a result of the separation of the Australian continent on the west and south. The destruction of the Old Plateau has involved etchplanation to the former weathering front and the evolution of a 'New Plateau' surface (Fig. 5). The New Plateau cycle is most advanced in the catchments of the Indian Ocean drainage in the west of the area, whereas the Old Plateau surface survives extensively on inland interfluves, now consisting largely of sandplains and areas of lateritic gravel. The Old Plateau remnants are bounded by long low 'breakaways' or escarpments leading down to the New Plateau and revealing weathered bedrock of the lateritic profiles (Fig. 4).

In the shield desert of central Australia, situated at the continental watershed, granite hills have remained prominent above the erosional plains of the Cenozoic land surface, and lateritic crusts have been generally confined to former valley tracts.

In both areas of shield desert, smooth erosional plains pass in turn into alluvial washplains of gentle gradient, traversed by shallow unchannelled floodways. The stability and age of these plains are reflected in their deeply leached red earth soils, indicative of pedogenesis under moister past conditions, and perhaps of the incorporation of pre-weathered sediments from the lateritic land surfaces upslope. These red-earth plains, subject to sheet flow, are characterized by the arrangement of shrubs in groves extending along the contour for a kilometer or so (Fig. 5).

On the washplains of Western Australia the red earth subsoils are cemented by a siliceous hardpan which may extend to within a few centimeters of the surface. It has been found that the hardpan is very close to the surface between the lines of vegetation, but drops sharply to form a narrow trench up to 1.5 m deep beneath the shrubland grove, presumably in consequence of localized infiltration [5].

With increasing distance from upland sources of streamfloods, washplains give way downslope to the sandplains of sand desert. However, on equivalent lower slopes on the shield desert plains of Western Australia, a more active outgoing drainage has interacted with the wind-transport of sand to form patterns of sandy banks alternating with alluvial flats known as *wanderrie country* [6]. The contours of the ever-present siliceous hardpan respond to the contrasts in soil texture and permeability, being deeper beneath the sandy banks and shallower beneath the finer-textured soils of the intervening flats.

The change from the moister climates of the earlier Cenozoic to the semi-arid

Fig. 4. Shield desert: 'breakaway' of ferruginized and mottled granite separating the 'Old Plateau' of interior Western Australia from the 'New Plateau', on which tors of unweathered rock have been etched from the ancient weathering profile. Photo by CSIRO.

Fig. 5. Shield desert: smooth washplains marked by groved shrubland are characteristic of the lowermost slopes of shield desert, as in central Australia.

conditions of the mid-Cenozoic led to the disorganization of river systems operating on the low gradients of the shield deserts. In Western Australia the rivers occupying the long shallow valleys broke up into aligned 'river lakes' which now serve as separate drainage terminals. Many of these valley and lake systems became filled with gravels and finer alluvium which were then cemented and capped by calcrete. Such carcreted valley trains now form low tabular rises of surface limestone with chalcedonic silica, known as 'opaline country'.

Stony desert

These landscapes, named for the widespread mantle of stones or 'gibbers', are best developed in the Lake Eyre Lowlands in the southwest of the Great Artesian Basin.

Upland forms in the stony deserts are tablelands and mesas with flat summits controlled by massive cappings of a silcrete duricrust, which commonly rests on weathered profiles in the soft claystones and lithic sandstones of the Basin (Fig. 7). These stony tablelands, with their bossy outcrops and large boulders of silcrete, are reminiscent of the desolate boulder hammadas of the Sahara. The boulder pavements are locally patterned by circular networks of stony 'gilgai', patterned ground formed by the heaving of underlying soils with expansive montmorillonitic clays.

Large areas of the stony tablelands possess a uniform mantle of relatively deep silty soil, which probably originated as wind-blown dust trapped by a surface stone layer. Commonly, the pavement of surface stone is only a few centimeters deep, and is underlain by a meter or more of stone-free soil. This has suggested an upward extrusion of soil from the soil profile, through the swelling of its montmorillonitic clays after wetting, probably as the aeolian dust mantle accumulated [7]. The aridity of these landscapes, as reflected in their sparse vegetation, is a function of the gentle slopes and stony surfaces which oppose the concentration of surface runoff, and the fine-textured, generally somewhat saline soils.

The silcrete cappings of the tablelands (Fig. 7) form remnants of an ancient land surface of low relief which in mid-Cenozoic times was thrown into a series of broad domes and depressions in the western part of the Great Artesian Basin. Some of these depressions, along the lines of older structures, have guided major rivers such as Coopers Creek, and have been zones of continuing alluvial deposition. On the uplifted areas, erosion has generally breached the arches of the domes, and so the uplands now survive mainly as cuestas along the flanks of the structures, with steep escarpments facing inwards over stony plains where the little-weathered underlying rocks have been exposed along the anticlinal axes.

The stony desert lowlands (Fig. 6) also generally bear a close cover of transported silcrete gibbers derived from the erosion and stripping of the silcrete crust,

Fig. 6. Stony desert: plains strewn with 'gibbers', fragments of a former siliceous capping are characteristic of much of the Interior Lowlands, on the weak rocks of the Great Artesian Basin. Photo by CSIRO.

Fig. 7. Stony desert: tablelands and mesas capped by silcrete duricrust above pallid weathered rock are also characteristic of stony desert, as in the southeast of the Northern Territory. Photo by CSIRO.

and resemble the regs of Saharan landscapes. There are three regional variants of stony desert lowlands, which reflect the differences in bedrock.

In the southeast of the Lake Eyre Lowlands are plains and undulating country with dense stone mantles above solonetzic soils. These lowlands exhibit patterned ground on a remarkable scale, with sorted stone steps following the contours, separated by vegetated flats with crabhole depressions. They are a form of desert gilgai, strongly modified by the topography [7]. To the north and northeast are undulating lowlands or 'rolling downs' with cracking red clay soils. These too show banded gilgai patterns with silcrete boulders. The third area of stony desert lowland is the covered karst plain of the Nullarbor, with its chaotic patterns of 'rises and corridors', collapse dolines and underground caverns. The stony surfaces here generally have extensive calcrete mantles.

Riverine desert and clay plains

While the term riverine desert could be applied to all major floodplains, only the largest occurrences have been mapped in Figure 1. Of these, the most important areas occur in the east and southeast areas of the arid zone. On the slopes of the Lake Eyre Lowlands, large river systems rise on the better watered, northeastern rim of the arid zone, and flood in most summers, towards the driest parts of Australia. Here they have been obstructed by subsidence and backtilting, and more recently by dunes and alluvial barriers. Consequently, wide plains of fine-textured alluvium have been formed, reflecting the shale parent rocks in source areas. These major river systems, such as Coopers Creek, occupy floodplains of low gradient some tens of kilometers wide. The main active channels are flanked by raised tracts of reddish silts and fine sands, and the channels themselves are commonly anastomosing, to give an interlacing pattern in which deeper channels are interconnected and traversed by braiding shallow floodways, active only at flood stages. This complex of channels has given rise to the name 'Channel Country' (Fig. 8). At various intervals the multiple channels connect in straight, deeply incised reaches, a kilometer or more long, which contain perennial waterholes or 'billabongs'.

A second large area of riverine desert occupies the southeast margin of arid Australia in the Murray Basin, and comprises the lower sector of the Riverine Plains and the tributary floodplain of the Darling River. Much of the Riverine Plain consists of alluvium deposited by ancient 'prior streams'. These were sinuous bedload channels indicative of flash flooding and coarse sediment transport, and the alluvium grades outwards from the sandy channel fills, through levee silts to backplain clays. Much of the Riverine Plains bear a mantle of wind-deposited clay (parna) derived from the west during past arid periods, when source-bordering dunes also grew from the northern channel banks. The active

Fig. 8. Riverine desert: anastomosing channels of Coopers Creek in the 'Channel Country' of southwest Queensland.

floodplains are slightly entrenched beneath these prior stream deposits and are occupied by meandering suspended-load channels. Large swamp basins occur in the lowest parts of the Plains. Near the northwest margin of the Plains, chains of large claypans mark the disorganization of tributary drainage during dry phases of the Pleistocene.

Bracketed with the riverine deserts are plains with residual clay soils formed on shale and limestone. These include the Barkly Tableland in the Northern Territory and the Winton-Blackall Downs in west-central Queensland.

Sand desert

This term embraces the dunefields and sandplains of arid Australia, roughly equal in extent, which together cover 2 million km^2 of the continent, or nearly 40% of the arid zone.

The dunefields occur in lowland basins where reasonably well-sorted sands derived from sedimentary rocks on the upland rims have been transported by ancient river systems that have since retracted or vanished due to increasing aridity. Subsequently, these alluvial sands were reworked and formed into dunes

Fig. 9. Sand desert: parallel sand ridges elongated with the dominant wind occupy more than 1 million km^2 of Australia. The dunes are mainly stabilized by vegetation, except for the crests of loose red-brown sand. A spacing of 400 m and a relief of 15 m, as shown here in the Simpson Desert, are characteristic. Photo by CSIRO.

by wind action. Sandplains occur mainly on the lower margins of shield desert or on ancient lateritic residual surfaces, and their sands, although in part also of alluvial origin, are derived principally from granitic rocks of the shield. A prominent grit fraction in these sands concentrates as a surface veneer which resists wind-sorting to any depth, and so they remain as level plains, occasionally with broad undulations trending with the dominant wind direction. The sand-plain soils are mainly clayey sands, with 10% clay at the surface increasing to 25% at a depth of about a meter, but at depth one may find a variety of horizons, including gravels or lime pans.

Among the dunefields, parallel longitudinal ridges of striking continuity and regularity dominate (Fig. 9). These constitute the major 'sand deserts' of Australia, particularly the Simpson, Great Sandy and Great Victoria Deserts. These dunes are generally about 15 m high with a spacing of 300–500 m, and sometimes continue unbroken for tens of kilometers. Locally the ridges converge and link in Y-junctions which invariably point down the direction of the dominant formative wind as indicated by the dune trend. In arid Australia as a whole, the dunes form an enormous arc open to the west, and have extended eastward in the south, northward through the Simpson Desert on the eastern side and westward in the north. This anti-clockwise pattern of extension is consistent with the generally anticyclonic circulation of winds across arid Australia.

With the existing moderate aridity, most sand surfaces in the Australian deserts are now stable under a close cover of vegetation, and the mobile sands are restricted to the dune crests or to areas of excessive sand supply. Belts of more markedly fossil dunes with rounded, well-vegetated crests and pronounced soil profiles occur on the perimeters of the arid zone, indicating a retraction of the arid core of the continent.

The interdune swales vary in character from catenary sandy surfaces near sand-source areas, where the dunes are generally lower and closely spaced, to flats with loamy soils in the more open, regular dune systems. Other variants include stony corridors where dunes have transgressed across stony plains and where they tend to open out and become more massive, and claypans and flood-outs where large rivers enter the dune systems and die out, notably the Finke River in the northwest of the Simpson Desert.

Desert lakes

The large number of 'lakes' that dot the map of arid Australia are an expression of the breakdown of connected surface drainage under aridity, in a setting of generally low relief. Closed lake basins without outflow can exist only where evaporation accounts for all surface inflow. Most closed desert lakes in Australia are very small, most of them playas with relatively short flooding cycles; like the feeder channels therefore, they tend to be activated by individual heavy rainfalls. For most of their existance these lakes present dry surfaces, and the features of these dry lake floors, as well as of their environs, are determined by their water, sediment and salt budgets, which in turn reflect relationships with the surface and groundwater systems.

The largest desert lakes occur in the bottoms of topographic and structural basins, where they also form groundwater terminals, as for example Lake Eyre (15,000 km^2) in the Great Artesian Basin. As in Lake Eyre, the form of the lake basin itself may have been determined partly by young tectonic movements and partly by deflation of lacustrine sediments. In these lakes the watertable generally lies close to the surface, serving as a baselevel for wind scour. The floors are dominated by evaporites from capillary solutions. The central part typically has a thin salt crust above saturated gypseous mud and lacustrine clay, which in the Australian desert lakes does not generally exceed a few meters in thickness, due to the combined influence of crustal stability and periodic deflation during dry periods. The evaporites may show a rough outward zonation, with sodium chloride in the central and lowest parts, followed by a gypseous crust, and finally an outward belt of lime-cemented shoreline deposits and deltaic flats. A marginal

Fig. 10. Desert lakes: with fringing sand dunes and salt crust.

zone of groundwater seepage, as at Lake Eyre, may be marked by soft saline muds, traversable only with difficulty, and the higher parts of such zones may also contain phreatophyte dunes. Along the western shores of Lake Eyre calcreted mound springs indicate an earlier period of stronger artesian discharge.

A larger number of smaller lakes occur at immediate levels on the slopes of structural and topographic basins, many of them enclosed by barriers such as dunes or river deposits. These lakes are generally by-passed by groundwater flow, and evaporites are likely to be less important in their floor deposits than the silts and clays deposited by entering river channels. The term 'claypan' aptly describes many such lakes. The floors of claypans are normally smoothed by deposition and planed by shallow lake waters driven by the wind, and generally present smooth surfaces with a mosaic of cracks. Parly saline and gypseous claypan surfaces generally become puffy on drying and the lake silts are then subject to deflation.

The open floors of desert lakes are important sites of wind erosion following the drying stage of a flood cycle, and leeward lake margins have in consequence, been areas of aeolian deposition. Where a source of sand exists, a large open lake floor represents an ideal surface for sand drift, and the wind-driven sand then accumulates on the vegetated lee shore as source-bordering dunes (Fig. 10). These are well-evidenced on the north shores of Lake Eyre and nearby lakes in the south of the Simpson Desert. Such dunes differ from those of the main sand deserts in their more active forms and less regular patterns, which may include nested parabolic dunes on the lee shores of large lakes.

Where claypans subject to alternate flooding and drying have had sufficient electrolytes to flocculate the silts and clays, the aggregates have been swept to the lee margin and formed into crescentic clay dunes or lunettes. These peculiarly Australian forms are common in a belt along the southern margin of the Australian arid zone where they are now fossil, reflecting late-Pleistocene drying episodes after a wetter period of greater lake activity. The lunettes have subsequently been leached of their original chlorides, the carbonates now form deep pans, while the main body remains a crusted mound of gypseous silty clay. Their age and fossil status are shown by the depth of leaching and pedogenesis, and today the lunettes are commonly subject to gullying and wind erosion. Erosion by end-currents commonly causes there claypans to become oriented with their long axes normal to the prevalent wind direction. Their lee shores, smoothed by wave action and aeolian deposition have a characteristic crescentic outline.

References

1. Mabbutt JA: Desert Landforms. Canberra, 1977.
2. Jennings JN, Mabbutt JA (1977): Physiographic outlines and regions. In: Jeans DN, (ed) Australia: a Geography. Sydney University Press, 1977, pp 38–52.
3. Mabbutt JA (1966): Landforms of the western Macdonnell Ranges. In: Dury GH (ed) Essays in Geomorphology. London, 1966, pp 83–119.
4. Twidale CR (1967): Hillslopes and pediments in the Flinders Ranges, South Australia. In: Jennings JN, Mabbutt JA (eds) Landform Studies from Australia and New Guinea. Canberra, 1967, pp 95–117.
5. Litchfield WH, Mabbutt JA: Hardpan in soils of semi-arid Western Australia. Journal of Soil Science (13):148–159, 1962.
6. Mabbutt JA: Wanderrie banks: micro-relief patterns in semiarid Western Australia. Bulletin of the Geol. Soc. Amer. (74):529–540, 1963.
7. Mabbutt JA: Pavements and patterned ground in the Australian stony deserts. Stuttgarter Geogr. Stud. (93):107–123, 1979.

Author's address:
School of Geography
University of New South Wales
PO Box 1, Kensington
NSW 2033, Australia

5. The sandy deserts and the gobi of China

Chao Sung-chiao

Abstract

Sandy deserts are widely distributed in northern and northwestern China, along with numerous gobi plains. (The term 'gobi' is a Mongolian word that denotes all the deserts and semi-deserts in the Mongolian Plateau. In China, the word 'gobi' is used to describe deserts and semi-deserts paved with gravel or rock debris, while 'shamo' is restricted to sandy deserts.) These deserts occupy a total area of about 1,095,000 km^2, of which about 637,000 km^2 are sandy desert and 458,000 km^2 are gobi. They account for 11.5% of the total land area of China. Although some of these lands are rich in resources, natural hazards such as drought and aeolian erosion frequently occur. This chapter is a study of the origin and evolution of the sandy deserts and the gobi in China, providing a basis for more detailed study.

Introduction

The Chinese deserts are located in a temperate zone, stretching from 75° E to 125° E, and from 35° N to 50° N. They are characterized by the following common features:

1) The climate is dry, with scanty and sporadic rainfall that decreases as the distance from the sea increases. From 106° E westward, annual rainfall drops below 200 mm, with a considerable part as low as 10 mm. Within the arid zone, from 106° E eastward, annual rainfall totals 200–400 mm, which is still inadequate for dry farming. Rainfall variabilities, both seasonal and annual, are very great; in parts of southeastern Xinjiang (Sinkiang), there might be no rainfall at all for more than three years.

2) The desert areas are locked in the inland basins or high plateaux, with elevations ranging between 500–1500 m. Ground surfaces are rather flat or undulating and are frequently swept by strong winds. The ground surface materials are generally loose and coarse, chiefly composed of sands in the sandy

El-Baz, F. (ed.), Deserts and arid lands. ISBN: 90-247-2850-9.

deserts and gravels in the gobi.

3) Under these climatic conditions, no perennial rivers fed by local runoff exist. There are a few larger rivers which originate in the surrounding high, snow-capped mountains that flow into the deserts. The distribution of ground water resources is varied and unbalanced: there are large amounts along the river channels and under certain piedmont plains, but small amounts in the rest of the region.

4) Soil profiles are poorly developed, with a low content of humus but a high content of soluble salts and carbonates. Locally, favorable conditions exist where large tracts of fertile arable lands are found. It is estimated that in northwest China (including Inner Mongolia, Ningxia, Gansu and Xinjiang) there are more than 260 million mows (1 hectare = 15 mows) of arable lands of which about 160 million mows have been cultivated.

5) Vegetation is sparse; in the desert and semi-desert zones, the vegetation cover is usually below 30% or even entirely absent in large tracts. In the steppe zone, the coverage is more extensive, generally more than 50%. It is estimated that there are more than 290 million mows of fixed and semi-fixed sandy deserts and gobi where there is sufficient vegetation for grazing.

6) Solar radiation is very high, yet the winter season is long and severe, with large annual and daily variations in temperature. Temperatures as high as 83° C (180° F) have been recorded in the Gurbantunggut Desert (45° N, 87° E).

Distribution of Chinese deserts

There are eight desert zones in China which can be grouped into three belts (Fig. 1 [1]).

Belt A

The desert belt extends from the Helan (Ho-lan) Shan (about 160° E) westward. About 90% of the sandy deserts and the gobi of China are distributed in this belt, including the numbered areas in Figure 1. These areas are:

1) The Tarim Basin (southwest Xinjiang). The sandy deserts and the gobi are arranged in concentric belts in an elliptic land-locked basin. The gobi, with an area of about 150 million mows lies on the rim of the basin, its inner side being interspersed with patchy oases and clayey flat lands, while the immense basin center is occupied by the largest sandy desert in China, the Taklimakan (Takla-makan: an Uygur word, meaning 'the place from which there is no return'), with an area of more than 500 million mows, about 85% of which is shifting sands (Fig. 2) [2, 3].

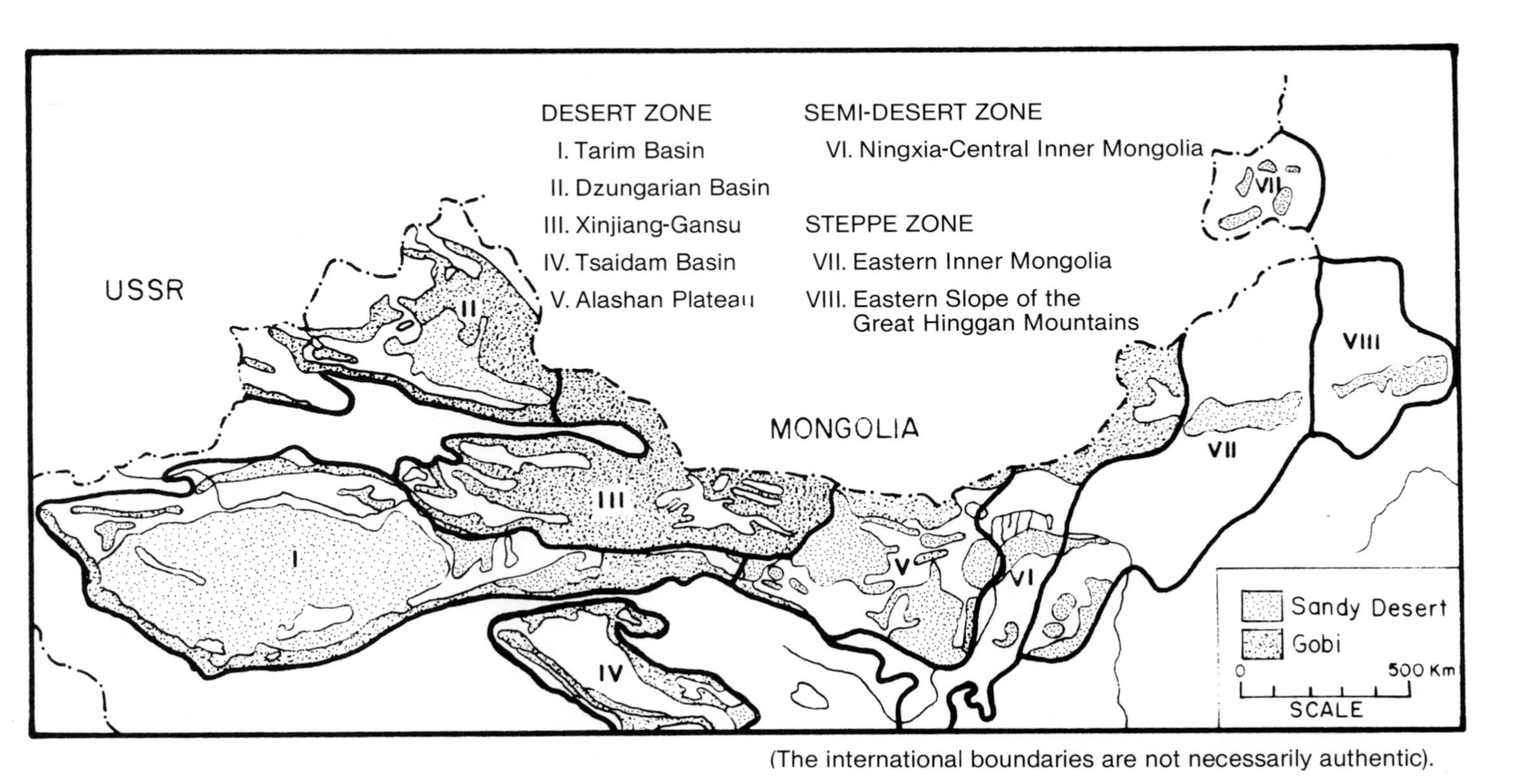

Fig. 1. A sketch map of the Chinese Deserts.

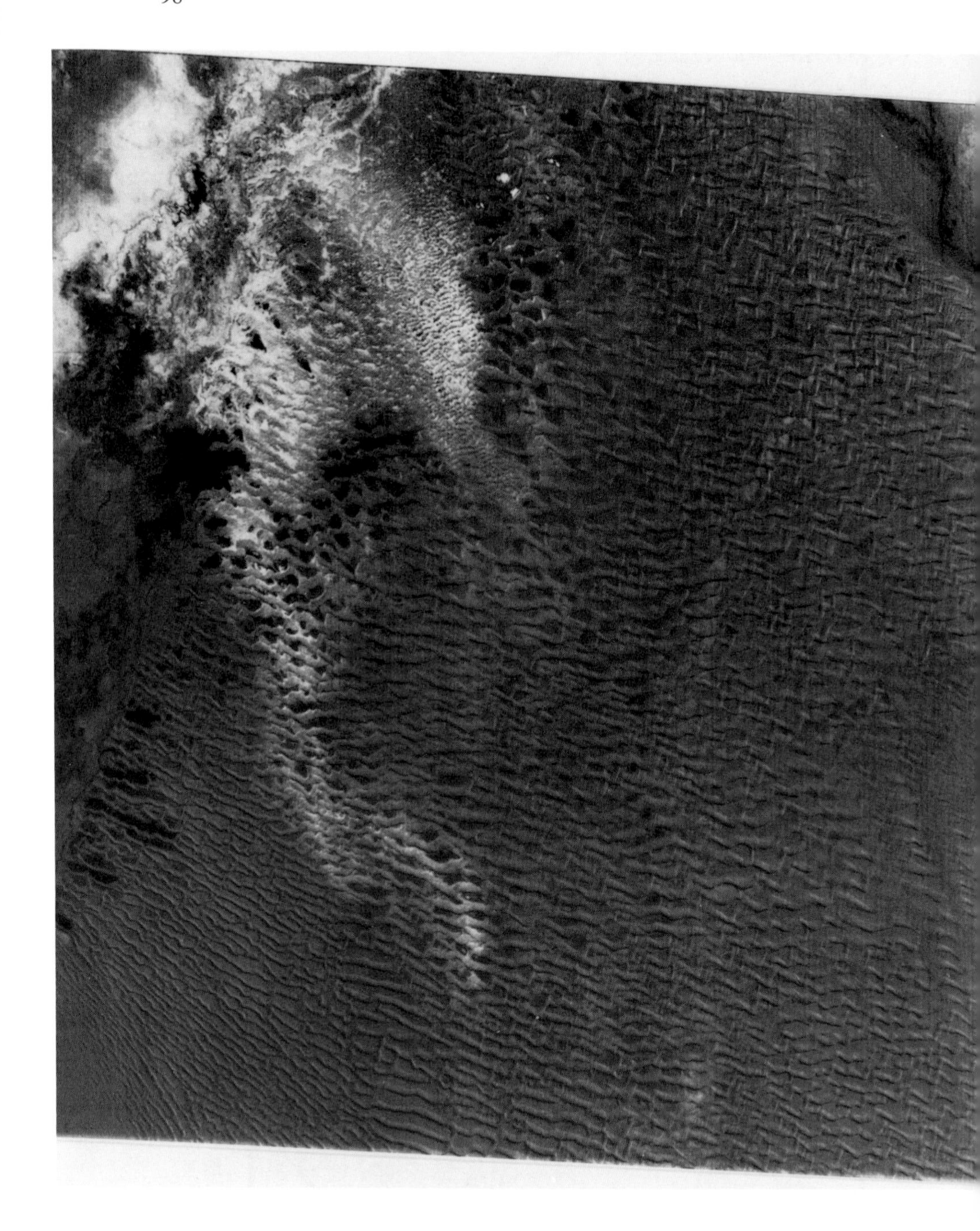

Fig. 2. Parallel dune chains in the northeastern part of the Taklimakan Desert, northwest China. In this Landsat image, the white areas in the upper right represent a light snow cover. The black spots are regions of collected water from snow-melting. The width of the image is approximately 180 km.

2) The Dzungarian (Junggar) Basin (north Xinjiang). The sandy deserts and the gobi are distributed in a triangular inland basin. The basin center is occupied by a vast sandy desert, the Gurbantunggut, with an area of about 90 million mows, of which only 3% are shifting sands. The gobi, with an area of about 110 million mows, also is found at the rim of the basin [4].

3) The Neighboring Region between Xinjiang and Gansu. This is mainly an intermontane area with different types of gobi clustered together. In the pene-planed Bei (Pei) Shan (Ma-chun Mountain), the desert varnish glitters vividly on the gravel surface and is thus called 'Black Gobi'. On the foothills of the Nan Shan (Ge-lin Mountain), the gravel deposit is thicker, the vegetation coverage more extensive, and no desert varnish exists; thus, the name 'White Gobi' [5].

4) The Quadam (Tsaidam) Basin (North Qinghai). Just as in the Tarim Basin, the sandy deserts and the gobi are arranged as concentric belts in an elliptic inland basin. However, here the sandy deserts are scattered, while in the basin center, there are mostly salt marshes. This is the highest desert zone in China due to its location along the northern margin of the Tibetan Plateau [6].

5) The Alashan Plateau. This is the southwestern part of the Mongolian Plateau (between 94° and 106° E). Different types of gobi are distributed extensively on this plateau. The sandy deserts are mainly located in three marginal areas: (a) The Badain Jaran (Badan-Karin) Desert (western Inner Mongolia, 41° N, 102° E) has an area of about 60 million mows and a close network of compound 'sand mounts' which sometimes surpass 420 m in height. These are probably the highest in the world. Among these sand mounts, there are many inland lakes [7, 8]; (b) The Tengger (Tengri) Desert (southwestern Inner Mongolia, 38° N, 104° E), with an area of about 45 million mows, is characterized by a mosaic of shifting sandy dunes and lake basins; and (c) The Ulan Buh Desert (west-central Inner Mongolia, 40° N, 105° E) is located between the Helan Shan, the Lang Shan and the Yellow River, and it occupies an area of about 21 million mows, of which 39% is shifting sandy dunes.

Belt B

The semi-desert belt stretches from the Helan Shan eastward to the Wontormiao (42° N, 113° E) Burlinmiao (Burline-miao) (41° N, 110° E) Otok (39° N, 107° E) Dingbian (Dinbin) (39° N, 107° E) line. In this zone, the sandy deserts the gobi are rather patchy. This zone includes only one region.

6) The Ningxia (Ninnsia) - Central Inner Mongolia region. The gobi is widely distributed on the Mongolian and Ordos Plateaus (about 39° N, 108° E). The sandy deserts are concentrated mainly in the northwestern margin of the Ordos Plateau, while a number of isolated shifting dunes are scattered on the Yellow River's valley floor.

Belt C

The steppic belt extends from the aforementioned line eastward to the eastern slopes of the Greater Hinggan (Khingan) Mountains (43°–51° N, 118°–124° E). By definition, there is no gobi in the steppe belt. The sandy lands are widely separated and mostly fixed by vegetation. There are two zones within this category.

7) The Eastern Inner Mongolia region. There are three important sandy patches: (a) The Mu Us sandy land is situated on the southeastern margin of the Ordos Plateau, with fixed sand, half-fixed sand and shifting sand each occupying about one-third of the total area. Large tracts of swampy lowlands lie between the sandy dunes; (b) the Onzin-dag sandy land (45° N, 113°–116° E) is located at the northern slopes of the In Shan; 98% of this area is comprised of fixed and half-fixed sands [9]; and (c) the Hulun-bir sandy land (48°–50° N, 116°–121° E) is the northeasternmost desert in China. It features three mostly fixed sandy areas along ancient and modern river channels.

8) The Eastern Slopes of the Greater Hinggan (eastern Inner Mongolia) Mountains. The sandy lands are mainly distributed along the Western Liao River with fixed sand dunes interspersed with swampy lowlands. It is the so-called Kolshin sandy land (42°–47° N, 120°–125° E). It has a better moisture condition and more intensive land utilization than all other sandy lands in China. Yet, owing to continuous misuse of this land in the past, about 10% of the land has been degraded into shifting sands [10].

Origin and evolution

The sandy desert and the gobi are a product of the dry climate. The latter is closely related to geographic location, atmospheric circulation, geomorphic features and other natural factors. For example, in the subtropic high-pressure belt, tradewinds with very low humidity prevail and the vertical atmospheric movement is mainly of subsidence type, resulting in very little precipitation. In the case of northwest China, the great distance from the sea coupled with the barrier effect of the Tibetan Plateau and a series of lofty mountains (Greater Hinggan Mountain, In Shan, Helan Shan, Gelin Mountain, Tian Shan, Kunlun Mountain, etc.) play an important role in the shaping of the arid environment.

According to recent paleo-geographical studies [11], the dry climate in northwest China was formed as early as late Cretaceous and early Tertiary. The Chinese mainland was then mostly under a subtropic high pressure belt, with northeasterly tradewinds prevailing, and the terrain peneplaned, so that a broad belt of arid zone with subtropic park savanna and semi-desert vegetation extended from northwest China southeastward to the lower Yangtze Valley.

However, it was not until the onset of the Himalayan tectonic movement in the middle Tertiary that the modern Chinese climatic regime started to take shape. Late in the Tertiary, the Chinese mainland emerged into a vast subcontinent, and the Tibetan Plateau was uplifted, so that the continentality of the climate was greatly strengthened and the monsoon system became well established. At that time, the climate in the Yangtze Valley became moist, whereas northwestern China became even more arid.

Based upon recent research by the scientific investigation team of the Tibetan Plateau, as well as upon many archaeological excavations in northwestern and northern China [12], the forming of the dry climate in these areas are clearly correlated with a rapid uplifting of the Tibetan Plateau. Fossils of a three-toed horse (Hipparion) of the late Pliocene period, recently excavated in Gyirong (about 29° N, 84° E) and other places in the Tibetan Plateau, are nearly analogous with those excavated in northern China. Both regions might have been of a subhumid foreststeppe environment at that time, while the northwest had a semi-arid steppic landscape. In late Pliocene, the Tibetan Plateau attained an elevation of about 1,000 m; the Siberia-Mongolian High Pressure was not yet formed, and there only existed a weak high pressure belt near Lhasa (about 30° N).

At the end of the Tertiary period, the Tibetan Plateau together with its neighboring regions was violently uplifted. The plateau surface attained an elevation of about 3,000 m. There no longer existed any forests on the plateau, while the weak high pressure belt near Lhasa was strengthened and pushed northward to about 40° N. Thus, desertification in northwestern China was greatly accelerated; the ancient lakes in the Tarim and other inland basins diminished or dried out gradually, and the Taklimakan and other sandy deserts probably enlarged considerably at that time. There still exist several ancient dry channels south of a modern network of braided channels in the middle and lower reaches of the Tarim River; the southernmost penetrated deeply into the Taklimakan Desert (about 80–100 km), possibly representing the ancient margin of the desert.

From late Pleistocene to early Holocene, the Tibetan Plateau and its neighboring regions again underwent violent uplifting. Since then, the Tibetan Plateau has attained an elevation of about 4,000 m, the present vast desert areas have been formed, and the present Chinese monsoon system has become well established. The winter high pressure center has been greatly strengthened and pushed northward to its present position (about 55° N). From this center, dry and cold winds blow outward in a circular pattern to the surrounding area. At about 97° E, they are blocked by the lofty Tibetan Plateau, resulting in two divergent wind systems: the northwestern and the northeastern. The former dominates most of China during the winter, while the latter prevails up to the southwestern margin of the Taklimakan Desert, where it is superseded once again by the northwestern (Fig. 3). Thus, it is dry and cold in northwestern and northern

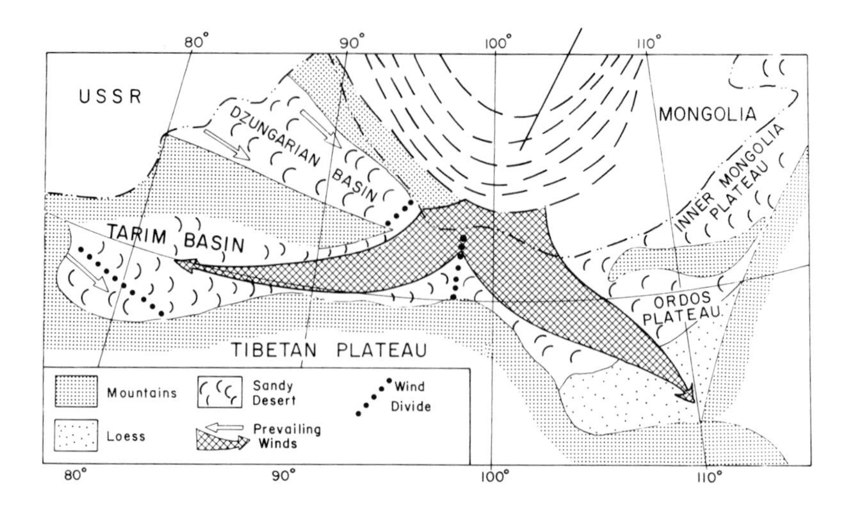

Fig. 3. Prevailing winds during the winter in Northwestern China.

China in the winter. During the summer, the southeastern monsoons predominate, bringing plenty of rainfall to southeastern China. However, in northwestern China and the western part of north China, due to the long distance from the sea as well as the barrier effect of the lofty mountains and plateaux, it becomes increasingly difficult for the moist maritime monsoons to penetrate westwardly, and hence, from the Greater Hinggan Mountains westward, with increasing distance from the sea, there appears a complete array of steppe, semi-desert and desert landscapes.

Recent numerical experiments [13] draw similar conclusions about wind directions (Fig. 4). The January geopotential map of the northern hemisphere at 1,000 millibars (unit of atmospheric pressure equal to 1/1000 bar, or 1,000 dynes per sqaure centimeter) including the effects of the Tibetan Plateau clearly identify the Siberia-Mongolian High at about 55° N, while a model not including the effects of the Tibetan Plateau has its weak high pressure belt at about 30° N.

Since the Holocene, the Tibetan Plateau and the surrounding mountains have been continually uplifting (in the Himalayas, the amplitude of uplifting might attain 13–14 cm/100 yr) so that the Siberia-Mongolian High and the Chinese Monsoon system have been continually strengthened. Thus, the arid climate in northwestern China has tended to be increasingly dry. Yet, from a human point of view, the rate of dessication is very low; its impact on the human society is negligible. Short-term climatic variations and abnormal climatic phenomena will exert a much greater influence upon human life. For example, the Hulun-bir sandy land is located in the transition belt between semi-arid and semi-humid climates, and thus is very sensitive to climatic changes [14]. The sand dunes are mainly concentrated in the semi-arid steppe zone, since their formation began in

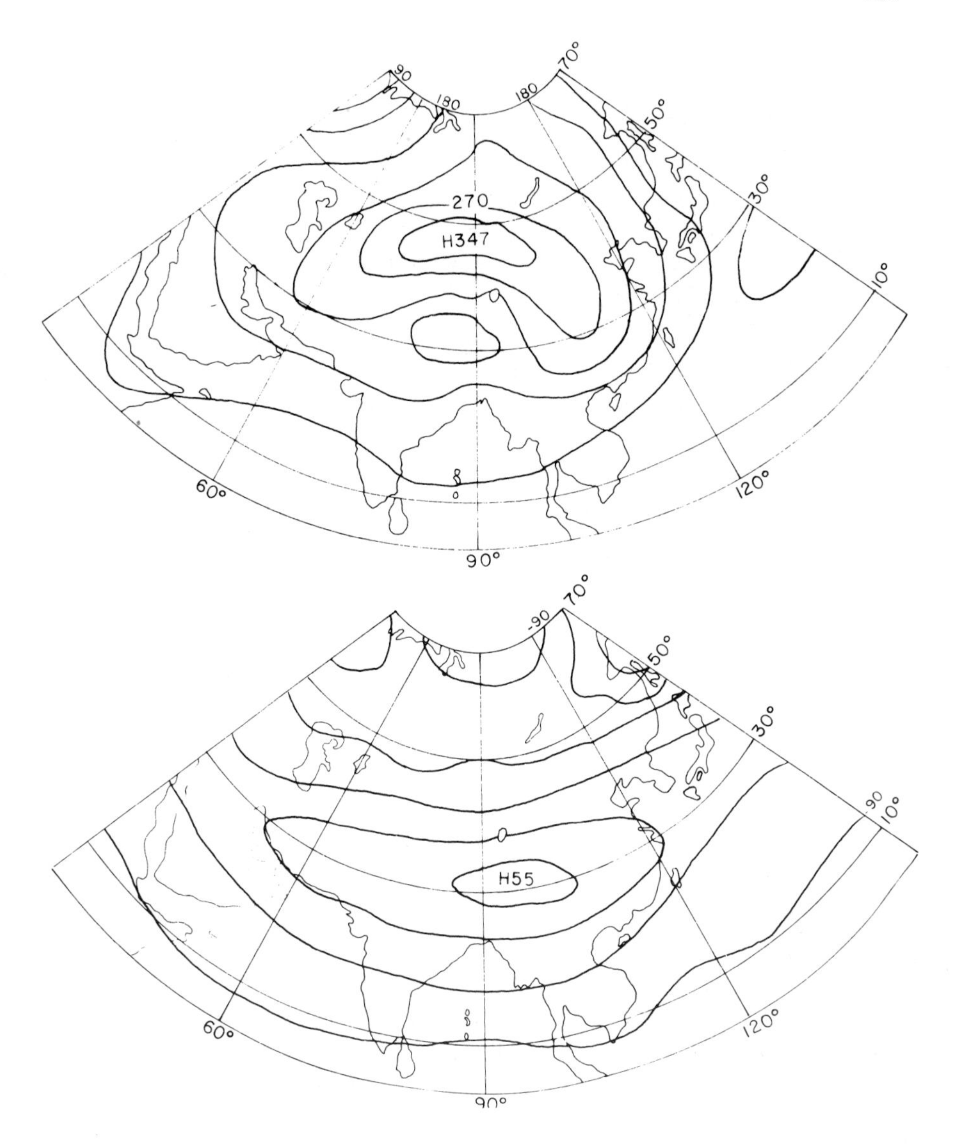

Fig. 4. January Geopotential map of the Northern Hemisphere at 1,000 mb: (top) With effects of the Tibetan Plateau; (bottom) Without effects of the Tibetan Plateau.

early Holocene. There are three basic layers in the otherwise yellowish dune profile of the western suburb of Hailar City, their depths being 55–95 cm, 260–330 cm and 425–525 cm, respectively. According to pollen analysis, these darker (blackish) layers contain fossil plants, which serve as an indication to the subhumid forest-steppe environment. Recently archaeologists in north and northwestern China, through excavation, carbon dating, and other techniques, have determined that the uppermost blackish layer might be from the transitional period between the neolithic and bronze ages (about 5000–3000 B.P.). It may therefore be tentatively concluded that since the early Holocene (about 10,000 B.P.) up to about 3,000 B.P., there were three cycles of alternating semi-arid and subhumid climates. Since then, the semi-steppe environment has prevailed again, with a newly deposited yellowish sandy layer about 50–100 cm thick now covering the land surface.

The climatic changes in recent centuries can be identified by dendrochronological data. For example, in the eastern part of the Hulun-bir sandy land, analysis of the annual rings of *Pinus sylvestris var, mongolica* revealed that the mean annual rainfall for the period 1790–1975 A.D. totaled about 360 mm, while instrumentally measured data for the 1961–1970 decade in Hailar city was 323 mm. Since 1790 A.D., there have been three drier periods, namely the early part of the 19th century, the 1850s through 1870s, and the 1910s through 1930s. Since 1960, another dry period has prevailed (Fig. 5).

Sources of desert materials

Under dry climatic conditions, the sandy desert and the gobi are formed through the integrated fluvial-aeolian process of erosion (denudation), transportation, and deposition. Numerous scholars have discussed such an integrated process. In this study only a few viewpoints will be mentioned.

Classification of landforms

Classification of the sandy deserts and the gobi types should be based upon integration of structures, processes and impacts of human activities.

The gobi can be divided into two genetic types: erosional (denudational) and depositional. These can also be subdivided again into denudational rock gobi, denudational-diluvial gravel gobi, erosional-diluvial gravel gobi, diluvial-alluvial gravel gobi, alluvial-diluvial sandy pebble gobi, etc. These gobi types and subtypes are distributed in sequential succession. For example, in the northwestern Hexi (Hosi) Corridor (39° N, 98°–102° N), where different types of gobi are closely arranged (Fig. 6), denudational rock gobi and denudational-diluvial gravel gobi are distributed on the peneplaned Bei Shan (41°–42° N, 94°–97° E) and its footslopes, while diluvial-alluvial gravel gobi are located at its southern pediment plain. From there to the Shule River valley, patches of alluvial-diluvial sandy pebble gobi are interspersed with oases and shifting sandy lands; again, in the northern slopes of the Nan Shan, diluvial-alluvial gravel gobi and erosional-

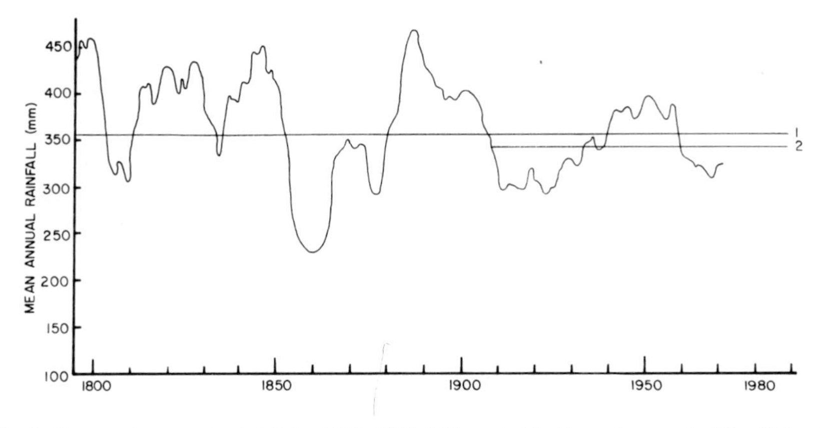

Fig. 5. Changes in annual rainfall for 1790–1915 A.D. period in the eastern part of the Hulun-bir sandy land.

diluvial gravel gobi appear once more.

There is still no universal standard used in the classification of sandy desert types. We believe that they can first be divided into three types: shifting (with vegetation coverage less than 10%), semi-fixed (with vegetative coverage 10–50%), and fixed (with vegetation coverage more than 50%). Again they can be subdivided according to the geomorphological characteristics of the underlying ground, e.g., sandy dunes deposited on alluvial plain, on diluvial plain, on hilly land, etc. The third category can be classified according to sand-dune types, such as barchan dune, linear dune, dome dune, etc. [15]. Different types of sandy deserts also have specific distributional patterns. The distribution of the first category of sandy desert types manifests the law of natural zonation: shifting sands in the desert zone, fixed and semi-fixed sands in the steppe zone, transitional conditions in the intermediate semi-desert zone. The second category of sandy desert types is arranged according to geological-geomorphological variations. The third category of sandy types is distributed as a function of prevailing wind direction, wind velocity, and underlying ground characteristics.

Origin of materials

The parent materials of the desert sands and the gobi gravels are mostly of local origin or from immediately neighboring areas. Water is usually the chief agent for the process of erosion-transportation-deposition, although wind can also be quite important and even dominant wherever there are strong and steady winds, such as in the Mongolian and Ordos plateaux.

Rock debris, gravels, and a small portion of sands on the erosional (denudational) gobi are the end-products of weathering and erosion (denudation).

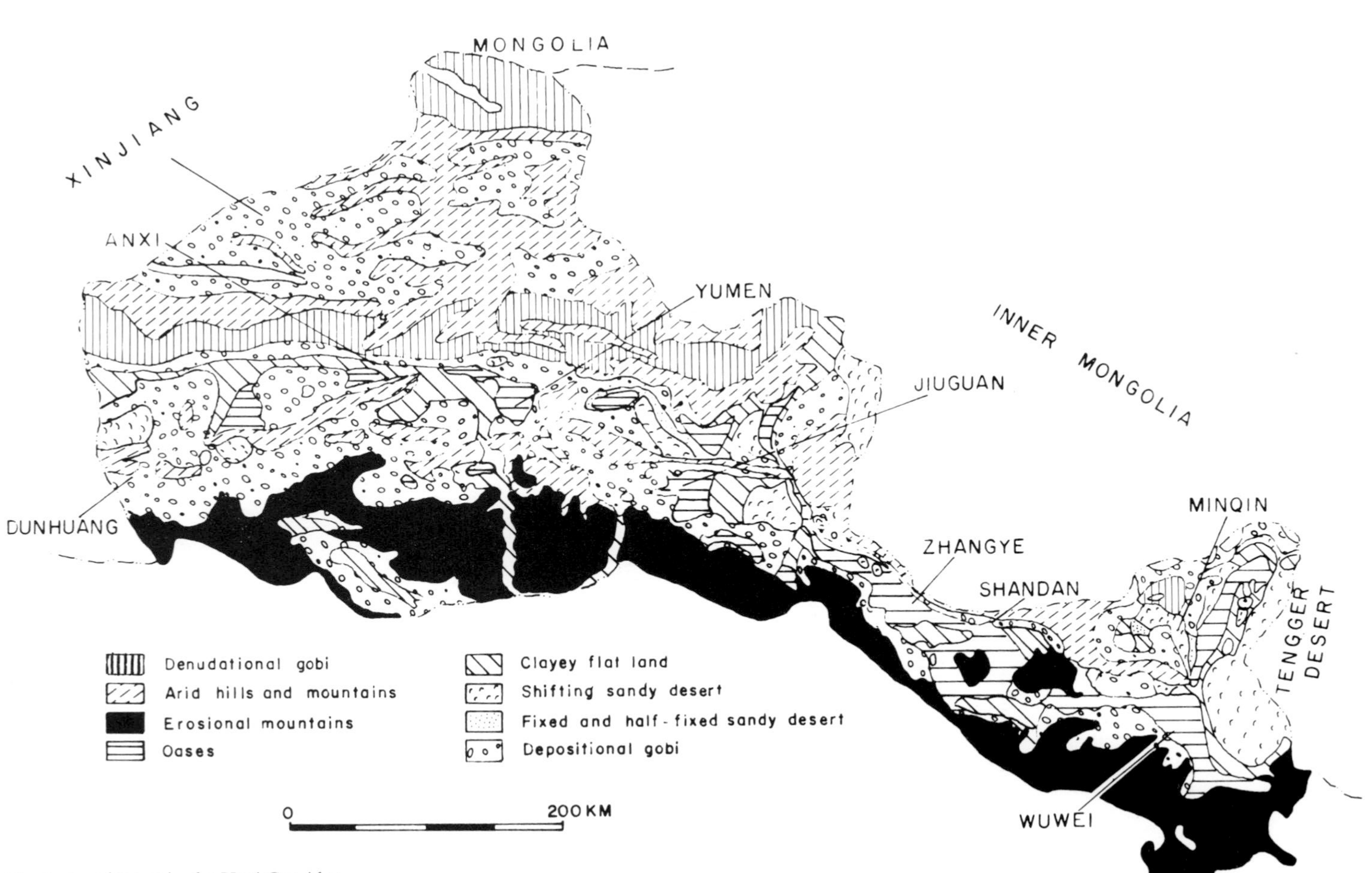

Fig. 6. Land types in the Hexi Corridor.

They are deposited *in situ* or transported a short distance through solifluction and diluviation; rock debris and gravel are generally angular in shape. On the depositional gobi, gravels are mainly transported and deposited by rivers or floods; they are rather circular or elliptic in shape. The winds, acting on these erosional (denudational) and depositional materials (by deflating, sifting, and sculpturing so that finer materials are selectively blown away) leave coarser materials behind, and eventually a barren 'gravel surface' is formed. For example, on the diluvial gravel gobi near Yumen, gravels more than 10–20 cm in diameter are strewn everywhere; very few fine materials remain. For the sake of gathering sands and other fine materials for engineering and agricultural uses, the local residents have to dig many pits on the gravel surface, in order to trap these fine materials which are carried in saltation and suspension along with the wind-sand stream. Such a gravel surface-forming process is one of the chief natural characteristics in the arid region. It begins to occur in the southwestern part of the Hulun-bir sandy land under a semi-arid climate. Here, owing to a large amount of fine materials having been deflated away, mainly small gravels and coarse sands remain on the ground, even though the surface is not yet paved with large gravels as under an arid climate. Hence, the Mongolian people give it the name 'tala', instead of 'gobi'.

The 'exogenetic' hypothesis for desert sands once was quite popular. Sven Hedin aptly concluded that the 'sands of the Taklimakan Desert have been blown in by the northeasters from the Lop Nur region (41° N, 91° E)' [16]. The Russian geologist V.M. Cinizin believed that the Taklimakan Desert appeared first in its eastern part and enlarged gradually westward as the dessication process accelerated [17]. Local residents in the Ulan Buh Desert also imagined that the sands in their desert had been blown in by the northwesters from the Badain Jaran Desert several hundred kilometers away. These viewpoints are now known to be in error.

The desert sands can be divided into two groups; aeolian sand (sandy-dune sand) and non-aeolian sand (underlying deposited sand). The latter are mainly transported and deposited through fluvial processes with circular or elliptical sand grains. Their parent materials are weathering residues and sediment materials in the same drainage basin. In regions with relatively abundant rainfall or with better developed drainage systems, they are particularly well developed with sandy lands concentrated along modern or ancient river channels. For example, in the Hulun-bir sandy land, the desert sands mainly come from the extensive sedimentary Hailar Layer (Q_3), and all major sandy belts are distributed along modern or ancient river channels.

In the Tarim Basin, huge amounts of weathering residues occur in the surrounding mountains, extensive diluvial and alluvial sedimental plains are stretched along footslopes, with deposits reaching 900 m thick, and in the basin center, there are vast alluvial plains which are composed of thick sediments deposited by ancient and modern rivers. All of these weathering residues and sediments, transported and deposited by the Tarim and other rivers, are a main source of non-aeolian sands in the Taklimakan Desert. Again, in the Ulan Buh Desert, non-aeolian sands come mainly from ancient alluvial deposits of the Yellow River in the north, extensive ancient lacustrine deposits in the west, and thin diluvial deposits along the footslopes of the surrounding hills.

Aeolian sands are mostly derived from the neighboring non-aeolian sands, deflated and deposited by winds. Their sand grains are often angular in shape and are composed chiefly of fine sand (0.25–0.1 mm in diameter) and medium sand (0.5–0.25 mm in diameter), the former occupying an increasing fraction as the age of deserts become older or prevailing winds stronger. The sands carried in the wind-sand stream are thickest in the near-ground layers, the lowest layer (0–10 cm above the ground) usually occupying more than 80% of total sands in the wind-sand stream. Layers more than 70–100 cm above the ground carry practically no sand at all, but do carry silts and clays with grain sizes less than 0.05 mm in diameter. The loess which is accumulated in the margin and outside the sandy deserts, generally has a granulometric composition of 0.05–0.01 mm in diameter.

Distribution model

The pattern of distribution of the sandy deserts and the gobi in China can thus be summed up into two models, with sub-models in each case.

The Inland Basin Model (Centripetal type) simulate conditions in the Tarim and other inland basins. From the surrounding mountains to the basin center, the sansy deserts and the gobi are arranged centripetally and concentrically. The fluvial process is more important than the aeolian process. The model proceeds as follows: montane terrain successively gives rise to erosional and depositional gobi; clayey flat lands and oases; fixed and semi-fixed sandy lands; and shifting sandy lands.

In the Qaidam Basin and the Hexi Corridor (Fig. 6), where the sand source is not abundant, salt marshes and clayey flat lands occupy the basin center, respectively. They can be termed as two variants of this model.

The Mongolian Plateau Model (Centrifugal type). From the plateau proper to the rimming mountains and surrounding regions, the gobi, the sandy deserts, and the loess deposits radiate from a central point. Both fluvial and aeolian processes are important. There are two sub-models representing the western arid region and the eastern semi-arid region respectively: (a) In the western arid region (desert and semi-desert zones) the denudational gobi gives way to shifting or half-fixed sand lands; rimming mountains; and loess deposits; (b) In the eastern semi-arid region (steppe zone), the 'tala' give way to half fixed and fixed sandy lands; rimming mountains; and loess deposits.

Environmental changes

As discussed above, the sandy deserts and the gobi in China have undergone significant natural evolution ever since their formation millions of years ago. Yet, compared with their environmental changes during human historical time, the rate of natural evolution has been rather slow.

The following cases are given as examples showing environmental changes in sandy deserts during historical time (the Gobi do not show conspicuous changes in historical time).

Arid regions

The natural conditions of sandy desert are rather severe, with natural hazards (drought, aeolian erosion and deposition, salinization, etc.) resulting in great damage. Key measures for transforming such a region include irrigation, sand control, and salinity melioration. These key measures usually account for the expanding or the diminishing of oases during historical time.

Bordering regions of the Taklimakan Desert. On bordering regions as well as in river valleys inside the Taklimakan Desert, irrigation agriculture existed as early as neolithic time. In the second century B.C. when Chang Chin visited the 'West Domain', he saw many cities and farmlands next to and inside the desert. Later, during the West Han, East Han, Tang, and Ching dynasties, irrigation agriculture was developed in these areas on a large scale. For example, during the reign of Han Cho Ti (86–74 B.C.) there were more than 500,000 mows of irrigated farmlands around Luntai (Lun-dai) and Yuli (Werli) cities in the lower reaches of the Tarim River. Yet, owing to natural or human reasons, there have been frequent changes in these oases and series of ancient ruins bear witness to these historical vicissitudes. Thus, the ancient extensive farmlands around Luntai and Yuli have now been all turned into shifting sands, and the famous ancient Loulan Kingdom, located at the western side of Lop Nur, has been utterly devastated into a wild landscape of yardang and barchan dunes.

The vast shifting sand dunes in the basin center, under the impact of prevailing winds (northwesterly in the southwestern part and northeasterly in all other parts), move slowly southward. Such a process has resulted in a slow southward extension of the Taklimakan Desert. According to recent investigations, many ancient castle ruins along the 'Silk Road' can be identified among sand dunes northward of the modern Pishan-Yecheng highway. Certainly, that busy and famous 'Silk Road' could not pass through the sandy desert one or two thousand years ago, but today it has been buried 3–10 km inside the sand desert.

On the other hand, farmlands have been increased rapidly on bordering regions of the Taklimakan Desert ever since the founding of the People's Republic of China. In the middle and lower reaches of the Tarim River, more than two million mows of new farmlands have been reclaimed, with clusters of new oases dotting the northern margin of the Taklimakan Desert (Fig. 7). However, owing to misuse or overuse of the land in local areas, patches of *Populous diversifolia* forests and productive meadows have been destroyed, resulting in an acceleration of aeolian hazards and salinization, both in the soil and in the water. These undesirable trends must be checked.

Northern region of the Ulan Buh Desert. According to archaeological and historical documents [18], human activities existed as early as neolithic time in the northern region of the Ulan Buh Desert. During the West Han dynasty, after defeating the Huns, the Su-fan prefecture was founded in 127 B.C. Among the ten counties of this prefecture, three were located in this region. Henceforth, for more than 300 years it served as one of the military farming centers in northwestern China. After 140 A.D. the farmers were compelled to evacuate the region, leading to the destruction of farmlands and irrigation systems, with aeolian erosion and deposition also greatly accelerated. Presently, ancient tombs from the West Han dynasty, which have protected the ground from aeolian erosion, stand one meter higher than the surrounding area, and large tracts of shifting sands are deposited wherever the overlying clayey layer has been eroded away.

Semi-arid regions

In primeval conditions, only a few patches of fixed and semi-fixed sandy lands existed in the semi-arid region, with negligible amounts of shifting sand dunes.

The Mu Us sandy land. This is the most notorious example of southward movement of the sandy desert in Chinese history [19]. Here, extensive lignitic strata are distributed in the depressions between sand dunes, sometimes attaining depths of several meters. On the tops of flat ridges, dry black soils are frequently found, while in the low lands and gentle lower slopes, small patches of forest thrive. One might conclude that this region was more humid during historical times, probably as a vast tract of grassland, spotted with a considerable amount of 'swamp jungle'. Traces of human activities can be found up to neolithic time, and their pattern of distribution clearly shows a temporal sequence. From the southeast to northwest, ruins of the West Han dynasty (206 B.C.–8 A.D.) extend deepest into the sandy lands, those of the Tang dynasty (618–907 A.D.) second, those of the Sung dynasty (960–1297 A.D.) third, while the Ming dynasty (1363–1644 A.D.) ruins are only distributed in the southeastern margin of the sandy lands.

Such a temporal sequence has been well correlated not only with spheres of

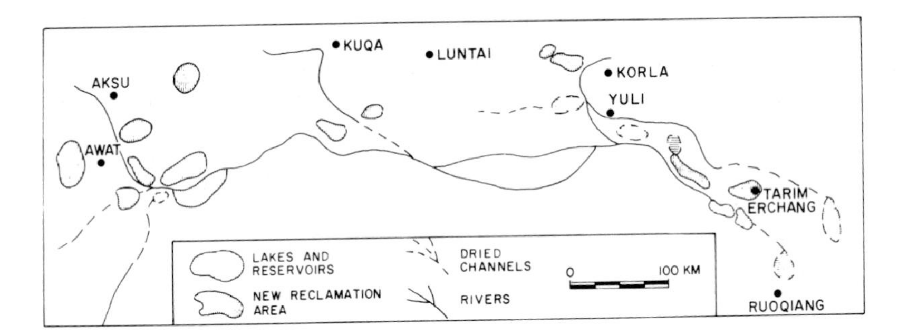

Fig. 7. New oases in the middle and lower reaches of the Tarim River.

political influences in different dynasties, but also with different stages of desertification. According to historical documents, the Mu Us sandy land was a fertile luxuriant grassland during the West Han dynasty, with prosperous agricultural and pastoral developments. In 413 A.D., the Shi Kingdom founded Tungfan city (the present 'White City' in northern Chenpian county (37° N, 109° E), Shaanxi Province) as its capital, with a population of about 200,000. Its suburbs, grasslands, and meadows were extensive and the rivers were clear at that time. It was as late as 828 A.D. that wind-shifted sands started to be deposited in the city's suburbs. When the city was finally destroyed in 994 A.D., it was described as already deep in the sandy desert.

In 1473 A.D., the Great Wall was built on the southern margin of the Mu Us sandy land. It served as the boundary between pastoral and agricultural areas at that time, while large patches of shifting sands appeared in the agricultural area. Again, in the middle 19th century, the Ching government attempted to open the Mongolian wasteland. Large tracts of sandy lands in the southeastern part of the Ordos Plateau were ruthlessly cultivated, resulting in a further devastation of grasslands and an extension of the shifting sand dunes. The boundary between pastoral and agricultural areas at that time moved northward to the present provincial boundary between Shaanxi and Inner Mongolia. The once prosperous Chenpian county turned into a rather barren country, with shifting sands, half-fixed sands, and saline lowland occupying more than 70% of the total area. Henceforth, coupled with the accelerating human activities, the process of desertification has been increasing critically.

The Kolshin sandy land. This realm enjoys the best natural conditions of all of the Chinese desert regions, with extensive meadows stretching in the lowlands. In the 10th century, the Liao Kingdom was established in this fertile region, engaging in both agriculture and animal husbandry. During the Yuan and Ming dynasties, there existed tall trees and luxuriant grasses. From the middle of the 10th century, until 1949, however, the sandy land has been subjected to rigorous cultivation and overgrazing, so that in and around the farmlands, pastures, and settlements, the 'white sand hills' have expanded from scattered spots to elongated belts, with shifting sands occupying about 10% of the total sandy area. In addition, the hazards of aeolian erosion and deposition have become very severe in and around the sandy area.

Since 1949, the Kolshin sandy land has become an important part of the 'Great Shelterbelt' of eastern Inner mongolia and western northeast China. A series of shelterbelt networks have been planted inside oases to protect farmlands and to fix scattered shifting sand dunes.

Conclusion

The widely distributed sandy deserts and the gobi in China have taken shape step by step ever since late Cretaceous and early Tertiary time. They have been formed under dry climatic conditions through an integrated fluvial-aeolian process of erosion (denudation), transportation, and deposition.

The recent climatic trend in the Chinese deserts tends to be a continuation of slow dessication, with minor fluctuations of wet and dry climates. Yet, under natural conditions, the rate of dessication is very slow and it might be assumed that within decades or hundreds of years, landforms, the climate and other inorganic natural factors will undergo no significant changes. During historical time, however, the Chinese desert region has been subjected to increasingly greater human activity and its environmental degradation has been considerable.

Acknowledgements

An earlier version of this paper was printed, courtesy of Dr. Harold Dregne, in the ICASALS (International Center for Arid and Semi-Arid Land Studies), Texas Tech University. Thanks are due to Dr. Farouk El-Baz for editing the manuscript and to his staff at the Smithsonian Institution's Center for Earth and Planetary Studies for preparing this paper for publication.

References

1. Sung-chiao Chao: The sandy deserts and the Gobi in China: Their natural features and their transformation. New Construct., (in Chinese), 1964.
2. Chen-da Chu: Natural features in the Tarim Basin Deserts, Geographic knowledge (in Chinese), 1960.

3. Integrated Investigation Team of Xinjiang 1978: Changes in natural conditions after large-scale reclamation in the Tarim Valley, (in Chinese), yet unpublished.
4. Zi-pin Chen: Natural features of the Gurbantunggut Deserts, Collective Papers in Geography 5, (in Chinese), 1963.
5. Sung-chiao Chao: A preliminary discussion on types of the Gobi in China, Selected papers of 1960 annual meeting of Chinese Geographical Soc. (in Chinese), 1960.
6. Shun-chiang Yang: Investigation on the Qaidam Basin Deserts, Research in desert transformation 3, (in Chinese), 1962.
7. Shio-chun Yu et al.: Investigation in the Gobi of western Inner Mongolia and the Badain Jaran Deserts, Research in desert transformation 3, (in Chinese), 1962.
8. Tung-mu Leu et al.: Origin of southeastern Badain Jaran Deserts and their transformation, Research in desert transformation 3, (in Chinese), 1962.
9. Shio-li Goa: Lake basins and aeolian landforms in eastern Tengger Basin, Research in desert transformation 4, (in Chinese), 1962.
10. Chin Lee: Origin and transformation of the Kolshin Sandy Land, Geography (in Chinese), 1964.
11. Compiling Committee on Physical Geography of China, Academia Sinica 1979: Palaeogeography (in Chinese), Science Press Beijing, 1982.
12. Chun-yu Lin et al.: Impact of uplifting of the Tibetan Plateau upon climate of China, Collective Papers of Symposium of Tibetan Plateau (in Chinese), in press.
13. Marabe S, Terotra TH: The effects of mountains on the general circulation of the atmosphere as identified by numerical experiments, J. of Atmospheric Sciences 31 (2), 1974.
14. Sung-chiao Chao: The moving Sand and the dust bowl in the Hulun-bir Sandy Land (in Chinese) 1977 in Press.
15. Lanzhou Institute of Desert Research: Map of distribution of Chinese deserts, 1 : 2,000,000, Shanghai: Chung-hwa Publisher (in Chinese), 1974.
16. Sven Hedin, et al. The Sino-Swedish Expedition's Reports (in English) 1928–35.
17. Cinizin VM: Quaternary history of the Tarim Basin, Geological Bulletin (in Russian), 1947.
18. Chin-ze Hou et al.: Archaeological investigation and change of geographical environment in the Ulan-buho Deserts, Archaeology (in Chinese), 1973.
19. Desert Transformation Team, Academia Sinica, 1975: Report of integrated investigation group of the Mu Us Sandy Land (in Chinese), in press.

Author's address:
Institute of Geography
Academia Sinica
Beijing, China

6. Analysis of desert terrain in China using Landsat imagery

Chao Sung-chiao

Abstract

In China, terrain classification began when civilization started, with early classification schemes recorded as early as 2,500 years ago. Terrain types are classified according to physical and environmental conditions, with the number of classes and subclasses of a given terrain being dictated by the scale of the map. In this paper, a classification is given, taking into account landscape criteria as well as well as environmental factors. Four examples are discussed: The Jiayuguan area in the middle of the Hexi Corridor, the Turpan Basin and its neighboring areas, the Minfeng area in the south central Tarim Basin, and the Taijnar Hu area in south central Qaidam Basin. This classification was determined by coupling field study results and image interpretation.

Introduction

The terms 'terrain' and 'land', describe the overal physical attributes of a section of the Earth's land surface, including climate, landforms, hydrography, soil, vegetation and other physical factors. This also includes past and present human activities and their impacts upon the natural environment.

The extremely arid desert zone in China is widely distributed westward from the Holan Shan (In Chinese the words, 'shan' denotes mountain; 'ho' means river; 'hu' means lake; and 'shamo' means sandy desert. In Mongolian, the word 'nor' means lake; and 'gobi' denotes the stony or gravel desert.) (about 106° E), to the Sino-Russian boundary, and northward from the Kunlun Shan (about 36° N) to the Sino-Mongolian boundary. This area of about 1.1 million km^2 occupies 11.5% of the total land area of China, yet it is inhabited by less than 2.0% of the total population. Four natural regions are identified in this vast zone: the Alashan Plateau Temperate Desert, the Junggar Basin Temperate Desert, the Tarim Basin Warm-temperate Desert, and the Qaidam High Plateau Desert. The two most common terrain types are: the shamo and the gobi, which account for

El-Baz, F. (ed.), Deserts and arid lands. ISBN: 90-247-2850-9.

about 30% and 25% of the total desert area, respectively.

Terrain classification

Ever since 2500 years ago, Chinese scholars have been analyzing the terrain of the whole country. In the ancient classic 'Chow Li', five major terrain types were recognized: forested mountain, hilly terrain, level plain, riverine low terrain and swampy depression. Such a scheme of classification was further developed by the 'Kwan Tze', a book written in the Warring States (475–221 B.C.), which divided the terrains into three first-level types: plain, hill and mountain. Based upon soil and ground surface material, these first types were subdivided into 25 second-level types. These are certainly the earliest terrain classification systems in the world. Henceforth, terrain classification and evaluation has continued to be of major scientific interest in ensuing dynasties, because such research work is closely related with agricultural production and land tax collection.

Since 1949, with accelerating advances in agricultural production and scientific research, a great deal of scientific study on terrain classification and evaluation has been conducted in different parts of China. More pronouncedly since 1978, with the study of the geographic environment as well as an evaluation of the land resources and their agricultural potential in China, an overall terrain classification was begun. Also a mapping program at the scale of 1:1,000,000 for the entire country, and 1:200,000 for the major provinces and regions, and at 1:50,000 for sampling areas has been undertaken.

Elements or facets can only be studied and mapped on very large scale maps usually larger than 1:25,000. Two divergent opinions prevail in this respect: one regards them simply as morphologic 'elements' or 'facets' in the second-level terrain types; another holds them as third-level or basic terrain types which are characterized by homogeneous soil species and vegetation.

Our approach to terrain classification is based on a combination of parametric and landscape criteria, sometimes genetic environmental factors are also considered. The philosophy to which we adhere is, that it is the integration of all physical attributes, or the 'landscape', which determines a real differentiation into a certain level terrain type; yet, if major diversifying factors in a certain level of terrain type are selected as parametric criteria, then, not only will it become more easily measurable and with greater precision, but also it will reflect the total 'landscape', more fully because landscapes usually change with major diversifying factors. For example, a change in macro-landform will certainly lead to great changes in temperature, moisture, soil, vegetation and other physical conditions.

In the desert zone of China, the major diversifying physical attributes in terrain analysis are the ground surface materials and the processes for their formation, although in the surrounding high mountains, vertical zonal differentiation is more important. Preliminary first and second-level terrain classification systems are proposed in Table 1 and a general physical profile is shown in Figure 1.

Analyses based on landsat images

Space images provide a new tool for the classification of terrain in China's desert zones. Four examples of terrain analysis based on a preliminary study of Landsat images are given in the following section, including (Fig. 2): (a) the Jiayuguan area in the middle of the Hexi Corridor; (b) the Turpan Basin and its neighboring area; (c) the Minfeng area in the south central Tarim Basin; and (d) the Taijnar Hu area in the south central Qaidam Basin.

Table 1. First- and second-level terrain classification system in the desert zone of China.

First-level terrain types		Second-level terrain types		
I.	Clay and silt level terrain	I.	1	Slightly saline clay and silt level terrain
		I.	2	Saline clay and silt level terrain
		I.	3	Takyrs
		I.	4	Salt marsh (a: old; b: new)
		I.	5	Yardans (a: strongly developed; b: weakly developed
II.	Sandy level terrain (Sandy desert or Shamo)	II.	1	Shifting sand
		II.	2	Semi-fixed sand
		II.	3	Fixed sand
III.	Stony, gravel level terrain (Gobi)	III.	1	Alluvial-diluvial sandy gravel gobi a: b: with overlying loess c: with overlying clay sand
		III.	2	Diluvial gravel gobi a: old b: new
		III.	3	Denundational stony gobi
IV.	Denundational mountain and hill	IV.	1	Montane desert
		IV.	2	Montane grassland
VI.	Nival alpine	VI.	1	Periglacial cushion vegetation
		VI.	2	Eternal snow
VII.	Oases (planted vegetation)	VII.	1	Irrigated oases
		VII.	2	Non-irrigated oases

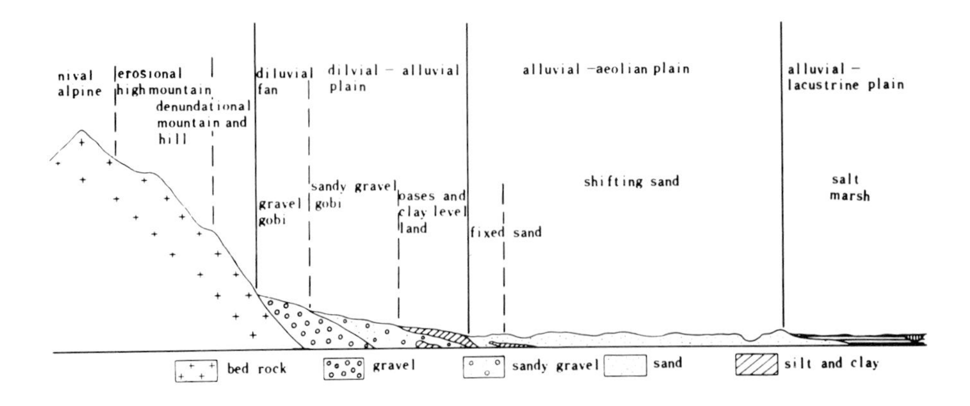

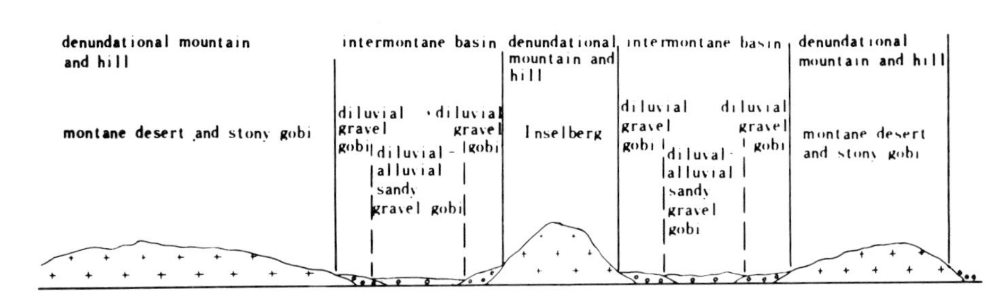

Fig. 1. Comprehensive physical profile of the Chinese desert zone: (a) inland basin type; (b) plateau type.

Jiayuguan area

Terrain types were studied on a Landsat image of the Jiayuguan area. This particular image was taken on 9 October 1975 (Fig. 3a). The Jiayuguan area is located in the middle of the Hexi Corridor on the southwestern Alaṣhan Plateau, sandwiched by the Qilian Shan on the south and the Machun Shan on the north (The Qilian Shan, owing to its geographic location, is also called Nan (South) Shan, while the Machun Shan, is also named Bei (North) Shan). The area is mainly composed of two basin plains: the Jiuquan Basin (a pre-Qilian Shan Piedmont plain) and the Jinta Basin (a marginal graben of the Alashan Plateau). The elevation descends from about 1800–2000 m along the southern rim of the Jiuquan Basin to about 1000–1200 m, along the northern rim of the Jinta Basin. Annual precipitation decreases correspondingly from about 80 mm to 50 mm.

The northern slope of the Qilian Shan generally has an elevation of 2000–4500 m, with its highest peak being 5934 m. The lowest part of the base (1800–2000 m) is a barren montane and hilly desert, with processes of denundation

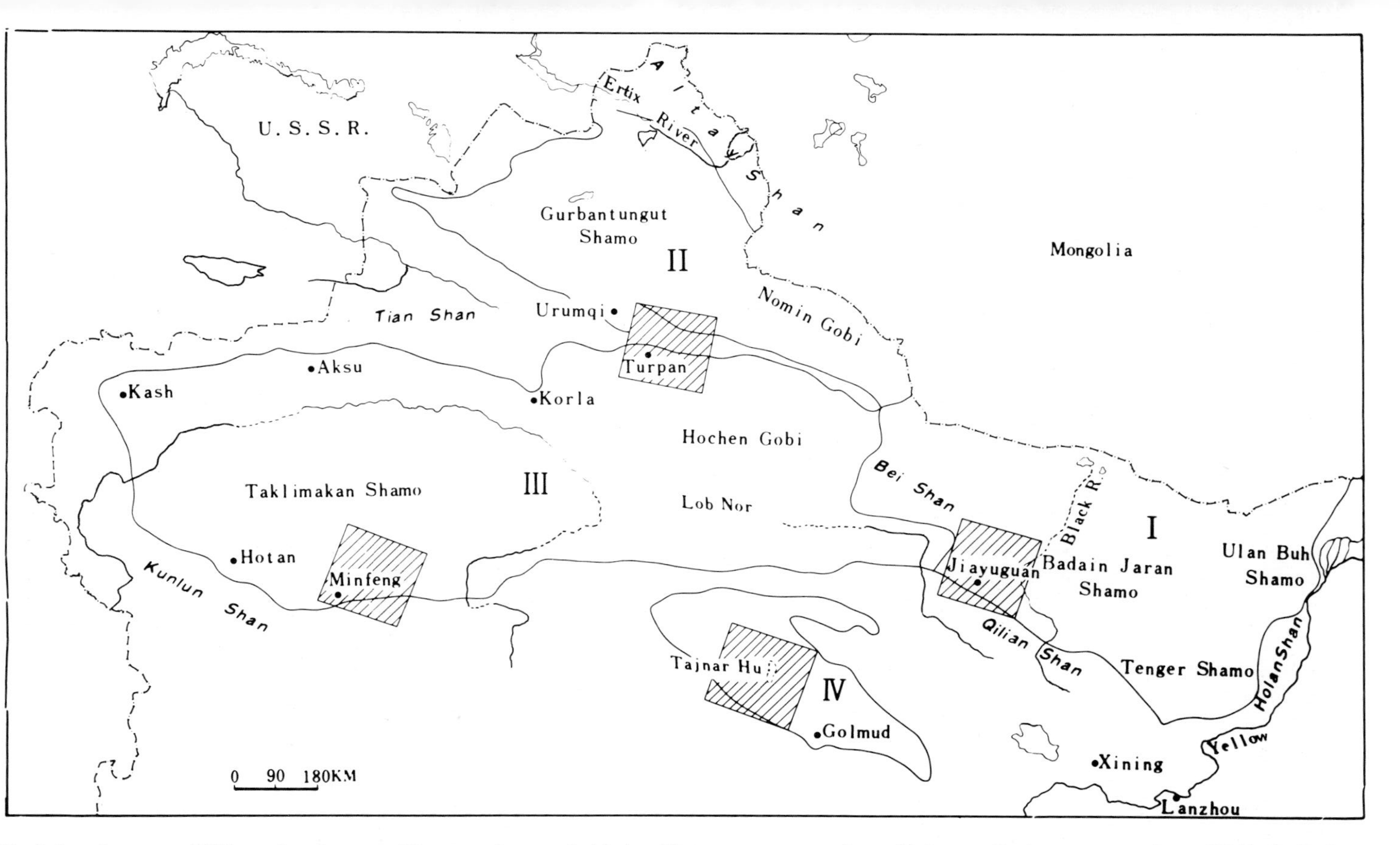

Fig. 2. Location map of Chinese desert zone and four sample areas: I: Alashan Plateau – temperate desert; II: Junggar Basin – temperate desert; III: Tarim Basin – temperate desert; IV: Quaidam Basin – high plateau desert.

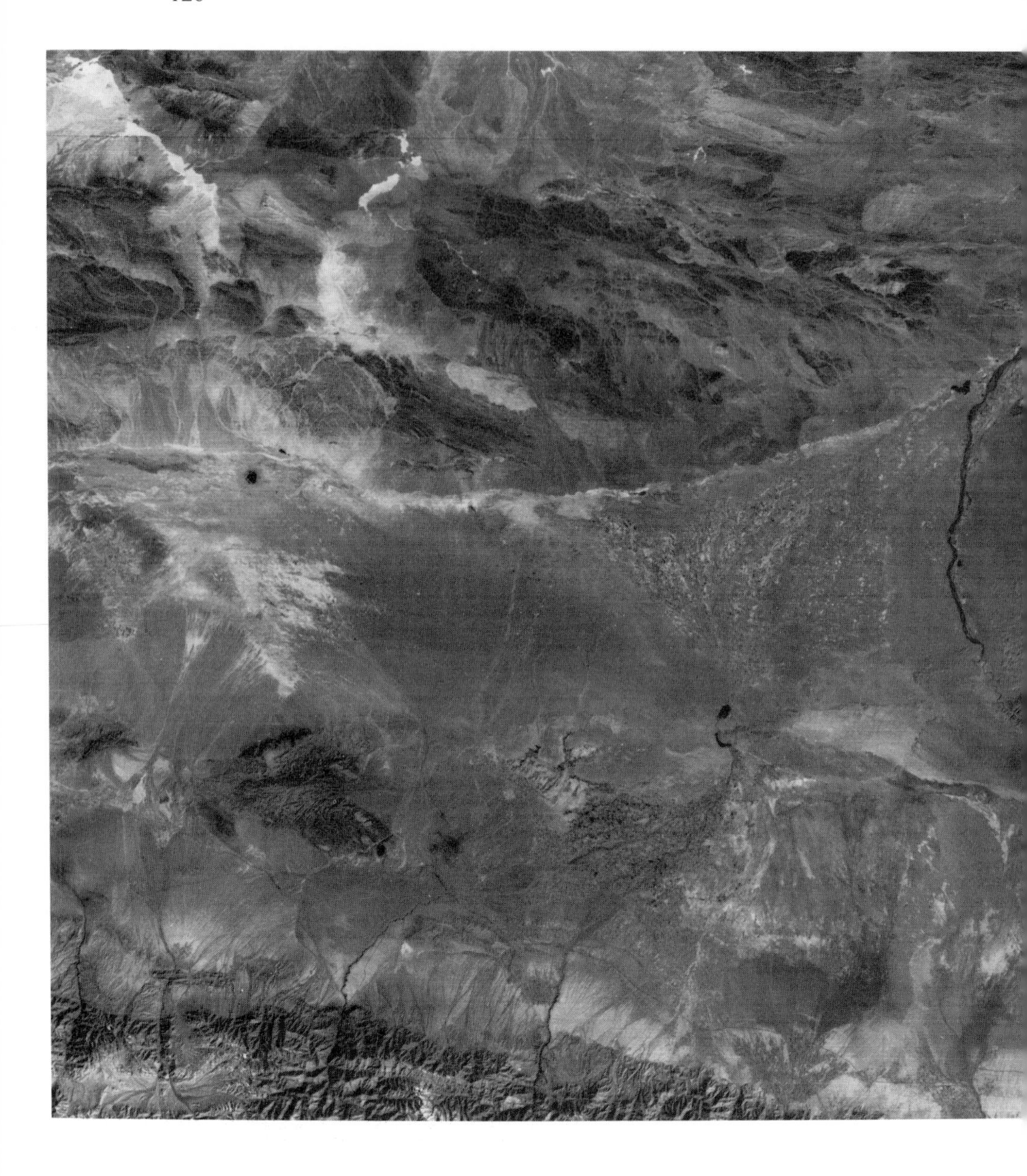

Fig. 3a. Landsat image of Jiayuguan area in the middle of the Hexi Corridor, China.
Fig. 3b. Drawing of Jiayuguan area showing terrain types in the area. Classification scheme is outlined in Table 1.
Fig. 3c. (A-A′) A comprehensive physical profile of the Jiayuguan area.

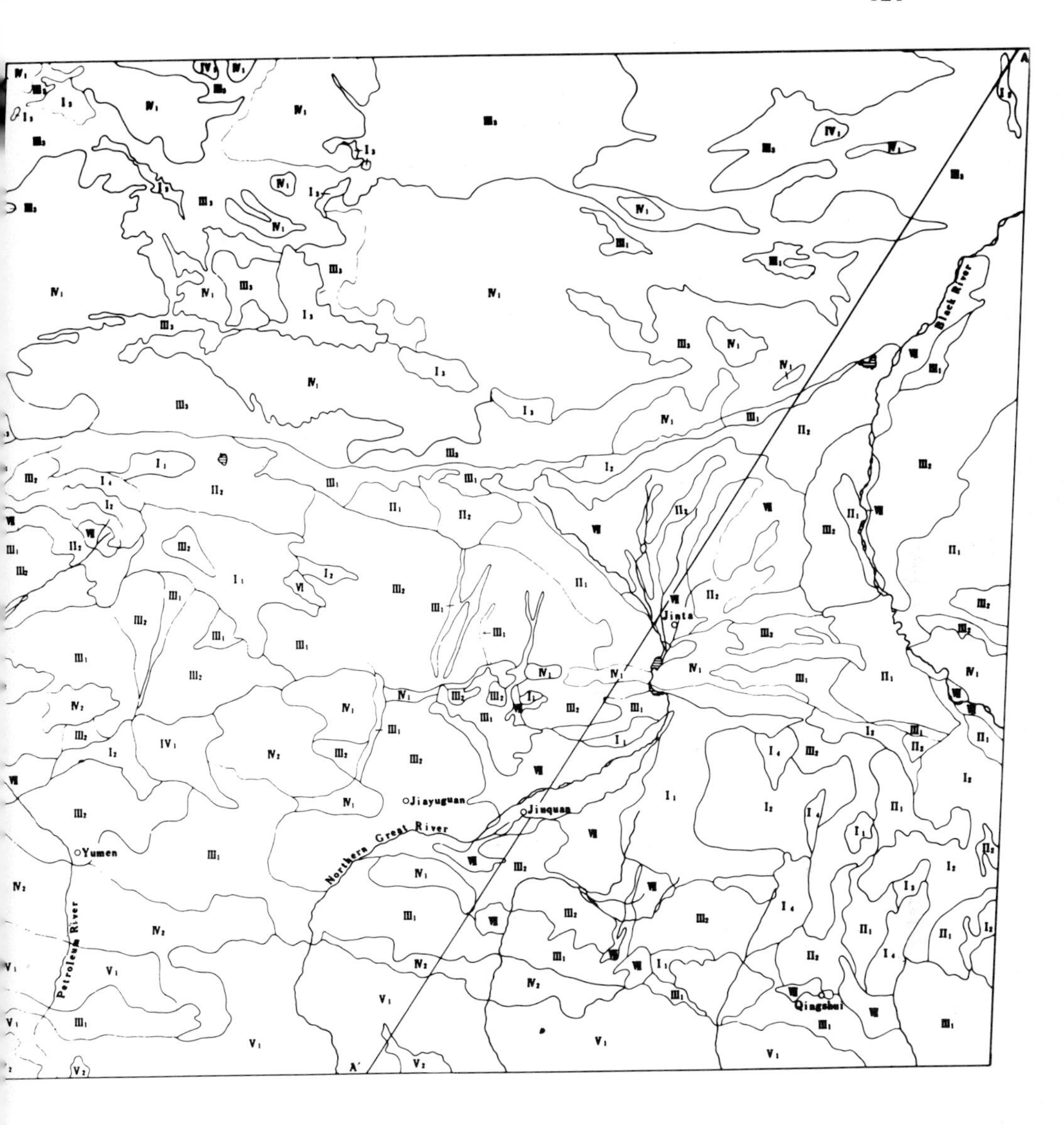
A
Black River
Jinta
Jiayuguan
Jiuquan
Northern Great River
Yumen
Petroleum River
Qingshui
A'

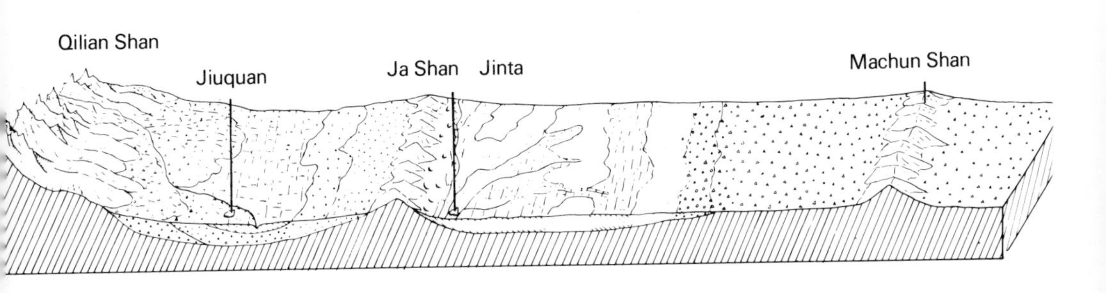
Qilian Shan
Jiuquan
Ja Shan
Jinta
Machun Shan

predominating. Between 2000 m and 3200 m, it changes into a montane grassland, composed chiefly of *Stipa*, *Artemisia* and other grasses, with active denundation and erosion. From 3200–3600 m, it is a montane coniferous forest belt, with *picea asperata* forest growing on the northern slope while *Stipa*, and *Artemisia* grasses grow on the southern slope. Above the tree line, up to about 4400 m, where the snow line lies, is found an alpine meadow belt, composed mainly of *Carex* and *Cobresia*, covering up to 80% of the land.

Further northward, the upper and middle parts of the diluvialalluvial piedmont plain are mainly occupied by gravel and sandy gravel gobi (Fig. 3b). Here, the ground surface slope is 1°–5° and the substrate of gravel and sand deposits reach several hundred meters in thickness. Vegetation is sparse, and the ground water table is deep. The lower part of the piedmont plain, however, has fine, fertile soils and bountiful water, and hence large patches of oases. The name for the ancient city Jiuquan literally denotes the 'spring of wine'; and the ancient castle of Jiayuquan was the last westward fortress of the Great Wall. There are also patches of saline clay and silt-level terrain and shifting sand. The Black Shan area, which is an uplifted block mountain that separates the Black River from the Shule Ho basins, is covered with denudational mountain and hill as well as gravel and sandy gravel gobi.

The Chia Shan is a W-E trending, strongly denuded frontal hill of the Bei Shan creating the marginal range of the Alashan Plateau. It attains only a few kilometers in width and a few dozen of meters in height, yet it creates a wide belt of hilly desertic and alluvial-diluvial sandy gravel gobi. It also creates a prominent george where the Great Northern River (a tributary of the Black River) cuts through it, providing the site of a new water reservoir.

Between the Chia Shan and the Machun Shan lies the Jinta Basin (Fig. 3c). It is a combination of large patches of oases, shifting and half-fixed sands and diluvial-alluvial sandy gravel gobi. There are also patches of saline clay and silt level terrain.

In the extremely arid Machun Shan area, the dominant terrain types are montane and hilly desert and denundational stony gobi. The darkish, glittering desert varnish is well developed on the surface of rock debris and gravels, and hence the name 'Black Gobi', that which strongly constrasts with the younger, depositional 'White Gobi' that developed on the piedmont plain of the Gilian Shan. In local intermontane basins of the Machun Shan area, there are also patches of younger, whitish sandy gravel gobi.

Turpan Basin

The Turpan Basin and its neighboring area are shown in a Landsat image taken 4 October 1972 (Fig. 4a), and in a terrain analysis based on that image (Fig. 4b).

This area is located astride the Bogda Shan (eastern Tian Shan), whose ridge stretches NWW-SEE with an elevation of about 4300 m. Consequently, this image covers three natural regions: the Tarim Basin in the south, the Tian Shan in the middle and the Junggar Basin in the north. The northern slope of the Bogda Shan is apparently more moist, a large area of which in eternal snow. The snow-line is found at an elevation of 3900–4000 m. The drainage system is also well developed, and a montane coniferous forest (strongly reddish in false-color composite) is distributed extensively just below the snow line. On a lower slope, some aeolian loess has been deposited, resulting in an abnormally high percentage of finer ground surface materials on the diluvial-alluvial piedmont plain, with extensive loess or clay beds overlying the gravel and sandy gravel gobi. Many springs empty onto the margin of the piedmont plain, creating large patches of oases and swamps. Further northward, the ancient alluvial-lacustrine plain in the Junggar Basin appears, with a cluster of oases that are interspersed with patches of whitish saline clay and silt level terrain.

The southern slope of the Bogda Shan is much more desiccated, with a smaller area in eternal snow and the snow-line located above 4200 m. Montane and hilly deserts and denunded stony gobi rise up to 2000 m, and a montane coniferous forest exists only on shady slopes. The process of desiccation is strong. Owing to frequent rainstorms during the summer, and a rich supply of rock debris, diluvial gravel gobi is well developed on the piedmont plain, with a width between 30 and 40 km; new layers often overlap older ones. In the older diluvial gravel gobi, desert varnish has already started to appear.

The E-W trending Turpan Basin structurally is a graben, sandwiched between the lofty Bogda Shan and the denunded, inconspicuous Garo Tagh ('Tagh' is an Uygur word, which is the equivalent of the Mongolian word 'Ula' denoting mountain and hill). The basin was initially formed during Mesozoic time, taking its present shape since the violent tectonic movement during the late Pliocene-early Quaternary era. It is again divided in two by a series of E-W trending frontal hills from the Bogda Shan, the Fire Shan, the Salt Shan, etc. The northern, or higher, intermontane basin is essentially the piedmont plain of the Bogda Shan, filled with old and new diluvial gravel gobi. Small streams disappear into the ground after flowing out of the Bogda Shan, only to reappear at the front of the diluvial gravel gobi, where patches of oases have been developed.

Similar lines are formed by the 'Karez', which are underground irrigation canals. The frontal hills generally have an elevation of 300–800 m, with a width of only 6–9 km. Their ridges have usually been peneplaned to form a belt of stony gobi. Several antecedent rivers cut through these frontal hills, forming gorges and ravines as deep as 150 m, the most famous of which is the productive Grape Ravine. The southern, or lower, intermontane basins are generally elevated below sea level (Turpan: −15 m; Toksun: −45 m; Ayding Hu, −155 m).

On the southern slope of the frontal hills, there again appears a belt of diluvial

Fig. 4a. Landsat image of Turpan Basin and neighboring areas in northwestern China.

gravel gobi. At its margin, is a clustered netwrk of rivers and karez as well as a series of fertile oases: Turpan, Toksun, etc. Further southward, a broad alluvial-lacustrine plain extends, dominated by clay and silt level terrain. The Ayding Hu is located at the basin center, with old and new salt marshes, surrounded by

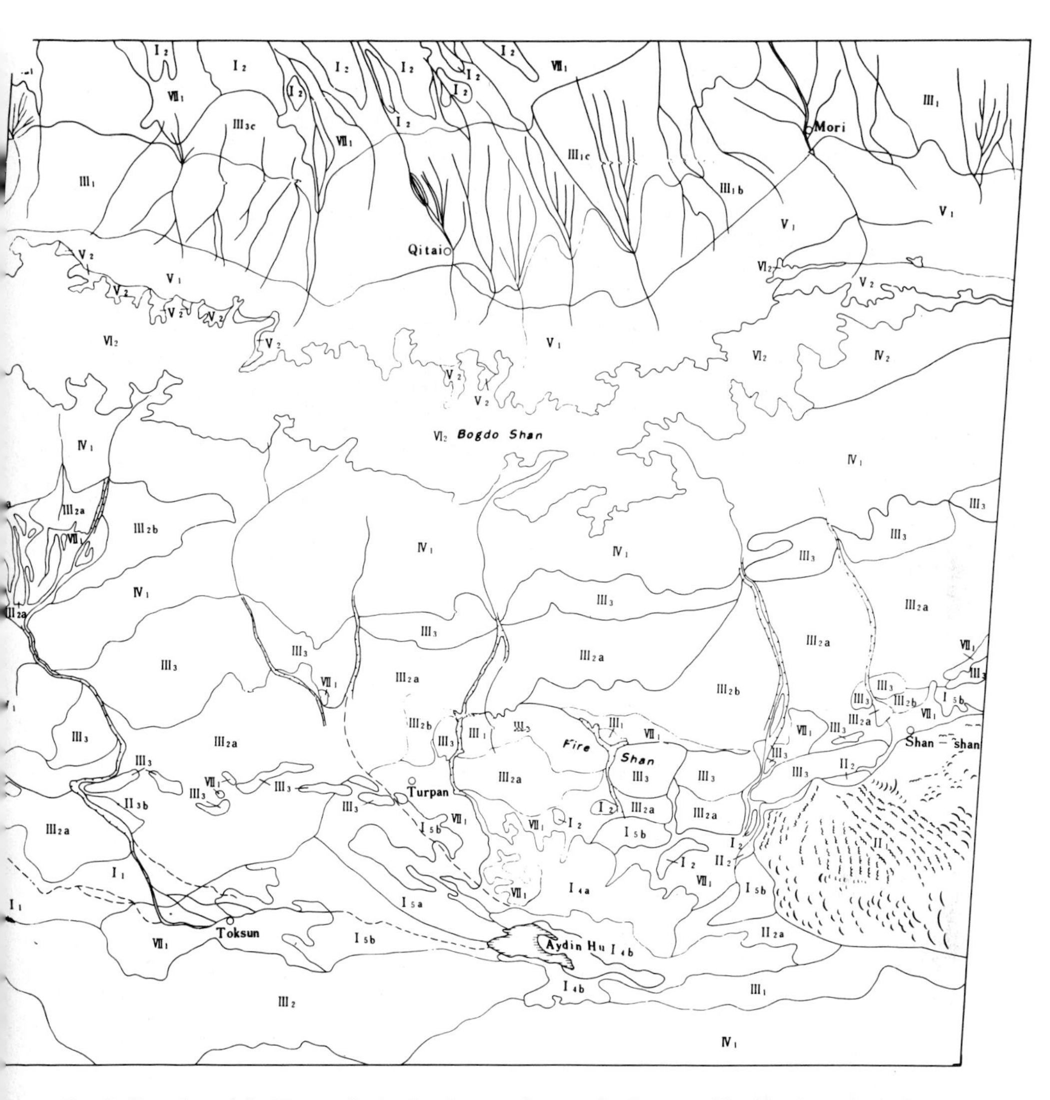

Fig. 4b. Drawing of the Turpan Basin showing terrain types in the area. Classification scheme is outlined in Table 1.

yardangs in varying stages of development. The lake itself nearly dries out during the summer when most river and karez water are used for irrigation. During the winter, however, its western part is filled by shallow saline water, while its eastern part remains as a salt marsh. Still further southward, there are once again, extensive diluvial gravel gobi, denundational stony gobi and the montane desert of the Garo Tagh area.

In the southeastern part of the Turpan Basin, another terrain type dominates, namely the shifting sand. It is developed on the peneplaned block tableland of the Kum Tagh. The sand dunes are formed by the interaction of two prevailing wind systems, the northwesterly and the northeasterly, which result in two major dune forms: longitudinal chains and pyramid dunes.

Minfeng area

The Minfeng area in the south central Tarim Basin is shown in a Landsat image (Fig. 5a), taken 30 November 1972, and an analysis of the terrain based on this image is illustrated in Figure 5b. This area is typical of the Tarim Basin region, with the lofty, arid Kunlun Shan overlooking a broad, barren belt of diluvial gravel gobi in the south, and the immense, Taklimakan Shamo stretching endlessly in the north. The region is located in the center of the largest continent in the world, Eurasia, with an annual precipitation of about 100 mm in the montane area, decreasing northward to less than 25 mm in the basin center. Here, the temperature is greater than, or equal to 10° C for a period of about 200 days of the year, with an accumulated temperature at that time of about 4000° C. Northeasterly winds dominate most of the region, yet from the Niya River westward of the Keriya River, there is a great transitional belt between the northeasterlies and the nortwesterlies; and, owing to the barrier action of the Kunlun and its frontal hills, there are also many local winds of different directions.

The lofty Kunlun Shan stretches southeastwardly from the Pamir Plateau as far as northwestern Sichuan Province, with a total length of more than 2500 km. The analyzed Landsat image only includes a very small part of the northern slope, with an elevation of 3000–5000 m (a nearby high peak is 6250 m). The barren, rugged montane and hilly desert extends up to about 3200 m. From 3200–4500 m, there is a montane grassland, very sparsely covered with dwarf graminious grasses and *Artemisia panvula*. From 4500–5000 m, there is a periglacial dwarf cushion vegetation, mainly composed of *Acantholimon diapesioides, Ardrosace squarrosula, Oxytropis poncinsu*, etc.

On the 30 to 50 km broad piedmont plain, as well as in some intermontane basins, there are some very extensive diluvial gravel gobi. The ground surface is paved with large, thick gravels. It dips from south to north, decreasing in elevation from about 3000 m to 1300 m. Rivers which originate in the Kunlun Shan also flow northwardly; most of them disappear in this coarse, porous area. Consequently, the ground water table is deep (usually more than 50 m), and vegetation is very sparse, with only a few shrubs scattered along wadis and river channels. Overlying gravel gobi, there are patches of shifting sand, composed mainly of small barchans and barchanic chains.

On the northern margin of the piedmont plain, many rivers reappear, the ground water table becomes shallower, and the soil becomes much more fertile. Here lie the Minfeng oases as well as many settlements and roads. Oases and sandy clay level terrains penetrate deeply northward into the Taklimakan Shamo along several large rivers, and then gradually disappear beneath the sand sea. During the geological and historical past, when the climate was somewhat more moist, and less water was used for irrigation in the upper and middle reaches, the 'green corridor' penetrated much deeper into the sandy desert. The ancient Keriya River, once flowed across the Taklimakan Shamo as does the modern Hotan River, to join the Tarim River. On the ancient Niya delta plain, more than 120 km north of the modern Minfeng (Niya) oasis, the famous Jin-Jua kingdom was founded during the Tang Dynasty (618–907 A.D.). Now, the ancient Niya ruins are surrounded by barchan dunes which are 3–5 m high, and along the lower Niya wadi stand rows and lines of dead *Populus diversifolia*.

Sandy desert covers most of the area in the Landsat image (Fig. 5a). Except for some strips of half-fixed and fixed sand dunes, which are located near oases and covered by rather dense stands of *Populus diversifolia* and *Nitraria spherocarpa*, shifting sands dominate the landscape. The sand is supplied by plentiful sand sources, namely deposits from runiing water, and then deflated and redeposited by blowing winds. The morphologic forms of the dunes are essentially determined by the direction and velocity of the wind. Three major dune types and seven subtypes are thus identified (fig. 5b):

1) *Transverse dunes:* with dune ridges more or less perpendicular to the prevailing winds. There are three subtypes: (a) barchan dunes and barchanic chains; (b) composite dune chains; and (c) chain-like sandy mountains.

2) *Longitudinal dunes:* with dune ridges more or less parallel to the prevailing winds. There are three subtypes: (d) barchan longitudinal dunes; (e) composite longitudinal dunes; and (f) composite longitudinal sandy mountains.

3) *Towering dunes:* formed by winds prevailing from differest directions, yet of similar velocity, meeting together, so that the sand dunes 'tower' upward. There is only one subtype in this Landsat image: (g) pyramid dunes.

Taijnar Hu area

The Taijnar Hu area in the south central Qaidam Basin is shown on a Landsat image (Fig. 6a) taken 29 November 1976, and the terrain analysis based on this image is shown in Figure 6b. The southern rim of the area is the northern slope of the Kunlun Shan, with an elevation of 3000–4500 m. It descends gradually northward through diluvial-alluvial plains and alluvial-lacustrine plains until it reaches the central structural depression of the Qaidam Basin the Eastern (Dong) and Western (Xi) Taijnar Hu, with an elevation of about 2700 m. At this point it gradually starts to ascend northward.

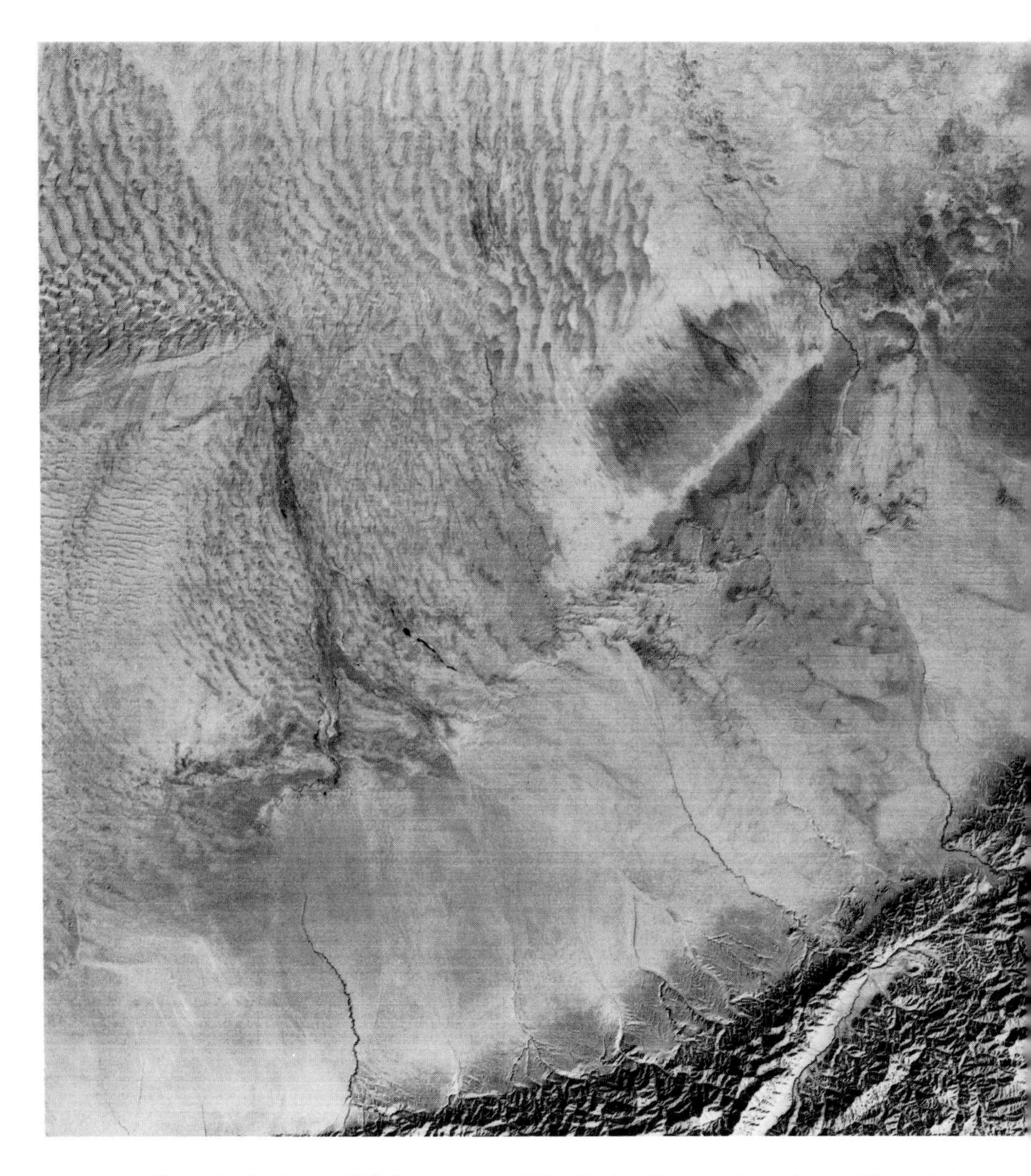

Fig. 5a. Landsat image of Minfeng area in south Taklimakan Desert in the south central Tarim Basin.

The northern slope of the Kunlun Shan in this region is essentially composed of pre-Mesozoic sedimentary and metamorphic rocks, strongly denunded and eroded, with steep slopes and rounded tops. The montane and hilly desert extends up to about 3600 m, with only a few *Ceratoides latens*, *Reaumuria soongarica*, *Ephedra przwalskii*, etc. covering the ground, Above 3600 m up to

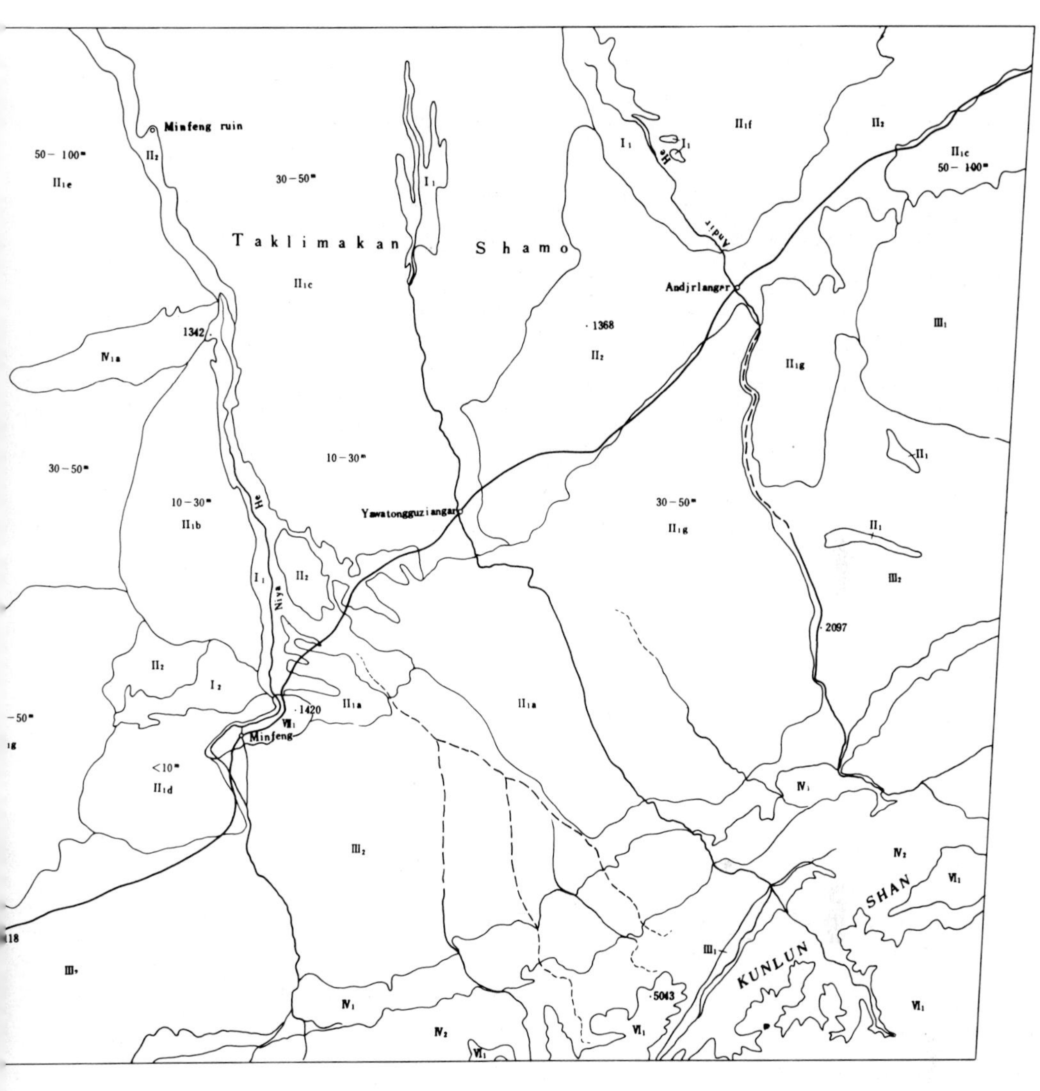

Fig. 5b. Drawing of Minfeng area showing terrain types characteristic of the area. Classification scheme is outlined in Table 1.

4500 m, montane grassland dominates, composed chiefly of *Stipa purpurea* and *Cobresia pygnaea*.

On the northern piedmont plain of the Kunlun Shan, diluvial gravel gobi is widely distributed. It generally has a width of 10–30 km and ground surface slope of 1°–4°. It is composed chiefly of Quaternary gravels interbedded with sands, the former being about 2–3 cm in diameter. It is nearly bare on its upper part, and very sparsely covered with shrubs and half-shrubs on its middle and lower parts.

Fig. 6a. Landsat image of Taijnar Hu area in Qaidam Basin, China.

Large patches of sand dunes, mostly shifting barchans, and partly fixed dunes, held in place by *Tamarix* and *Nitraria*, overlie and are interspersed with the gravel gobi.

From the piedmont plain northward, it reaches the lowest point of the Qaidam

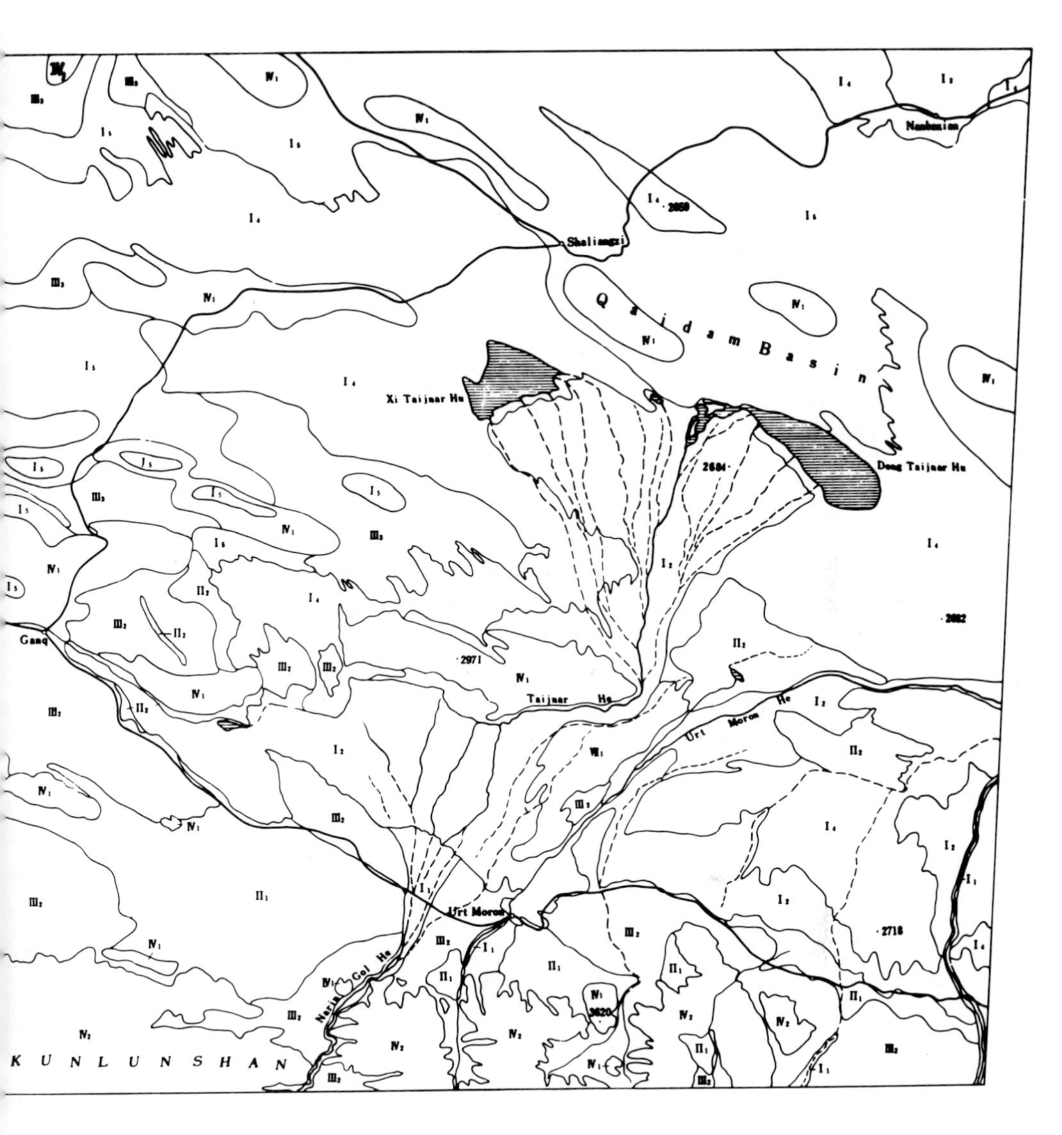

Fig. 6b. Drawing of Taijnar Hu area showing terrain types. Classification scheme is outlined in Table 1.

Basin, which is alluvial and alluvia-lacustrine plains. Except for one patch (a productive oasis which is located between two rivers and grows spring wheat, barley, irish potato etc. once a year), the surface is covered by overwhelmingly saline clay, silt level terrain and a salt marsh. The former is rather densely covered by saline meadow, composed chiefly of *Phragmites communis*, *Apocynum venetum*, *Carex* etc. The salt marsh usually has a salt crust of more than 60 cm, and a salt content of more than 60% in the upper soil layer. Hence it is nearly void of vegetation, except for a few *Salicornia herbacea*, *Suaeda salsa*, etc., which are scattered very sparsely on its margin.

The northwestern part of this Landsat image (Fig. 6a) is characterized by a series of NW-SE trending Pliocene to Pleistocene age anticlines, composed mainly of siltstone, sandstone and shale. In this region strong west winds blow during most of the year. Consequently, the most spectacular terrain type here is a series of extensively distributed yardangs with several meters width and few meters height, with some reaching up to tens of kilometers in length. This is the largest yardang field in China.

Conclusion

The desert is widely distributed in China, with an area of about 1.9 million km^2. As a whole, it is chiefly characterized by an extremely arid climate, and extensive plateau and inland basin landforms, as well as a scarcity of water and biological resources. In classifying the terrain, the chief parameters are landforms and ground surface materials. From the four examples of terrain analysis based on Landsat imagery, it seems that good and quick results can be obtained for mapping in the desert zone of China. The results are obtained when interpretations of the space photographs and remotely-sensed data are simultaneously coupled with the traditional ground investigations.

Author's address:
Institute of Geography
Academia Sinica
Beijing, China

7. Aeolian landforms in the Taklimakan Desert

Zhu Zhenda

Introduction

The Taklimakan Desert, the largest desert in the People's Republic of China (covering an area of 337,600 km^2) is found in the center of the Tarim Basin. The following is an attempt at explaining the origin of the sand, the morphological characteristics of the dunes, and their distribution and movement in this desert.

The origin of the sand of the desert

The origin of the sand of this desert is closely related to the Quaternary Paleogeographical environment of the subsurface relief underlying the sand dunes. With the exception of Mazar Tagh and Ros Tagh and the upland of north Niya (Minfeng), the greater part of the initial surface of the Taklimakan Desert consists of a series of diluvial-alluvial fans and alluvial plains. In the north is the alluvial plain of the Tarim River, the paleoriverbed of which stretches into the desert for about 80 km south of the present river course. To the east there are alluvial plains which were formed by the lower courses of the Tarim River and the Konche Darya, and the lacustrine plain of the Lop Nor. The western part is occupied by the alluvial plains of Kashgar Darya and Yarkand Darya, while the southern part is composed of the diluvial-alluvial fans and the alluvial plains built by the streams descending from the northern slope of the Kuen Lun.

The heavy minerals of the dune-sand are compositionally similar to those of the underlying deposits. The result of this analysis of the heavy minerals is that they can be classified into five groups: the southern part of the desert, where hornblende is the dominant mineral; the northern part of the desert comprised predominantly of metallic minerals and epidote; the southwestern part of the desert, where mica and garnet are dominant, and the eastern part of the desert where mica and hornblende are the chief minerals. A mechanical analysis also indicates a close relationship between the aeolian sand and the underlying deposits. For example, materials from the initial land surface in the southwestern

El-Baz, F. (ed.), Deserts and arid lands. ISBN: 90-247-2850-9.

part of the desert are very fine. The dune-sand appears to be the same size, with an average grain diameter of only 0.06–0.10 mm. The average grain size of the deposits from a different part of the same river varies as well, due to the different grain size of the underlying deposits. The case of the river Keriya may also be taken as an example. After descending the mountain, grain diameters of the river deposits range from 0.09–0.145 mm in the upper part, and are around 0.96 mm in the lower part. The above analysis indicates that the sands of the Taklimakan Desert are not carried by the wind from a region of wind erosion on the eastern border of the desert, but rather their origin was in the weathered materials of the bedrock derived from the bordering mountains. These materials, after being transported by water and deposited in the basin, have been remodelled by wind action.

Factors related to the development of the sand dunes

As mentioned above, the Taklimakan Desert is spread out across alluvial and lacustrine plains. In the arid climate, wind is the chief element responsible for the modelling of the surface relief. The role of the wind is manifested in two ways: Firstly, the wind direction determines the alignment of the sand dunes. In the Taklimakan Desert, the northeasterly and the northwesterly are the most important winds in the development of the sand dunes. Their dividing line is from Faizabad (Kia Shi) to Niya (Minfeng). Northeast of this line, the northeasterly prevails, and as a result the sand dunes trend from the northwest to the southeast. Southwest of this line, the dominant wind is northwest, and sand dunes trend in the northeast to southwest direction. Where the two wind systems meet, i.e. in the Faizabad, Niya and Keriya regions, the alignment of the sand dunes appears random. Within the Taklimakan Desert, although the alignment of the sand dunes shows obvious regional regularity owing to the influence of the two dominant wind systems, there are places where the alignment of the sand dunes trends in the east-west direction, due to local wind from the valleys of Tien Shan. In this case, the influence of local topography prevails.

Secondly, the forms of the sand dunes vary according to the different wind regimes. Under the influence of two winds blowing from opposite directions but with equal force, transverse dunes with symmetrical slopes will be produced. When the two winds cross obliquely at a lower angle, linear or seif dunes with one long wing will generally be formed. Under a uni-directional wind, regularly aligned barchanic chains will result. When two winds with equal force cross at right angles, longitudinal dunes will be produced. When winds from three or more directions meet, horn (cone) shaped dunes will be formed. Therefore, in the eastern and western part of the Talimakan Desert where a unidirectional wind prevails, barchanic chains of regular alignment are predominent. In the south and central parts of the desert, where wind directions vary, longiudinal and horn-shaped dunes are the dominant features. These examples are sufficient to show that a close relationship exists between the wind and the morphology of sand dunes.

Although wind is the basic dynamic element modelling the surface relief of the sand desert, various natural conditions on the sand exert considerable influence, so that the forms of the sand dunes also vary under the same wind regime. Where the supply of sand is plentiful, the dunes are high, where the supply of sand is insufficient, the dunes are low. In the center of the western part of the desert, where hilly ranges act as barriers to the movement of the sand-blowing wind, a great volume of sand accumulates on the windward slope and forms huge sand dunes 50–100 m high. Nearer the hilly ranges, the relative height of the sand dunes also increases, sometimes to over 100 m. From the north to the south of the desert there is a notable transition to imbricated transverse dunes, and then to complex transverse dunes, with sand dunes imbricated on the windward slope. The form of the sand dunes becomes more complicated as it approaches the mountain. Variation of the relief forces the air current near the ground to deflect. As a result, orientation of the alignment of the sand dunes also changes. For example, in the south central part of the desert, due to the existence of a highland running east to west, the northeasterly wind changes its direction to easterly when it encounters the north slope of the highland. This causes the sand dunes to change from northwest-southeast trending to north-south. Another example of the influence of local relief on the development of sand dunes is provided by the following case. The protruding relief itself may become the base of great sand dunes, forming dunes with extraordinary height, surpassing all other surrounding dunes. In the northwest part of the desert, some dunes reach a height of 200–300 m. This is associated with the existence of residual hills.

Water conditions and vegetation also exert influence on the development of sand dunes. In the north and south parts of the desert, the front zone of the diluvial-alluvial fan is a region with a relatively high ground watertable, with a depth of 1–3 m. During the summer and autumn months it is supplemented by flooding, and therefore vegetation is in better condition. Hence, sand dunes are represented by the fixed and semi-fixed dunes blanketed by shrubs of Tamarix. This is also true along the river which stretches into the desert. Consequently, the fixed and semi-fixed sand dunes of the Taklimakan Desert are distributed either along the river banks, or along the front of the diluvial-alluvial fan.

From this it follows that the formation and development of the desert landforms of the Taklimakan Desert are the result of the interaction between the wind and the sandy land surface in the arid climate, influenced substantially by the relief, water conditions, vegetation and the volume of sand. Along the fringe of the desert and around oases however, the development of sand dunes is closely related with human activity. In these places, destruction of the scrub on the fixed and semi-fixed sand dunes may bring about blown sand; deflation of the wind on the sandy surface where vegetation coverage has been destroyed may be the cause for the accumulation of sand. Sand dunes may also be formed by the changed course of a river which left behind a dry sandy riverbed, and afterward the action of the wind caused drifting of the river sand. According to a preliminary estimate, sand dunes of this origin occupy only 1% of the total area of the Taklimakan Desert.

Morphological characteristics of the sand dunes

In the Taklimakan Desert, besides the most common sand dunes, namely longitudinal dunes, barchan dunes, and transverse dunes, there are also sand dunes of a more complex type which are widely distributed, occupying nearly one half of the whole desert.

1) Complex barchan dunes (Fig. 1), chiefly distributed as single dunes, average 20–40 m in height, appearing as perfect barchans on the plane, imbricated with barchan dunes on the windward slope, with low leeward slopes.

2) Complex transverse dunes (Fig. 2) are one of the most widely distributed sand dunes in Taklimakan Desert. The length of each dune ranges from 5–15 km, and can reach up to 30 km; the width ranges from 1,000–1,500 m and the height from 80–200 m. The slipface is high and steep. The windward slope is densely distributed with transverse dunes. The whole dunal complex is at right angles with the direction of the prevailing wind. Between two dunes is found a depression.

3) Complex longitudinal dunes (Fig. 3) are also one of the most widely distributed types in the Taklimakan Desert. Their chief characteristics are: a great length averaging 10–20 km, but sometimes reaching 45 km; the alignment of the dunes is parallel to the prevailing wind; with a symmetrical slope. The surface of the slope is imbricated with sand dunes. Their height averages 50–80 m, and their width averages 0.5–1 km.

4) Horn (cone) shaped dunes (Fig. 4) are characterized by: trigonal slopes, inclined from 25–30 degrees with a pointed peak and sharp ridges, generally attaining 50–100 m in height.

5) Scaled dune groupings (Fig. 5) are characterized by: dense distribution in groups, with small interdunal spaces. The foot of the windward slope of the first becomes the base of the leeward slope of the second. When viewed from the individual dune, it stands at right angles with the prevailing wind. The wings of the dune stretch in the same direction as the prevailing wind and are thus connected with the windward slope of the next dune, forming a sand bar following the direction of the prevailing wind. Taking the whole group as a unit, it exhibits some characteristics of a longitudinal sand dune which is parallel with

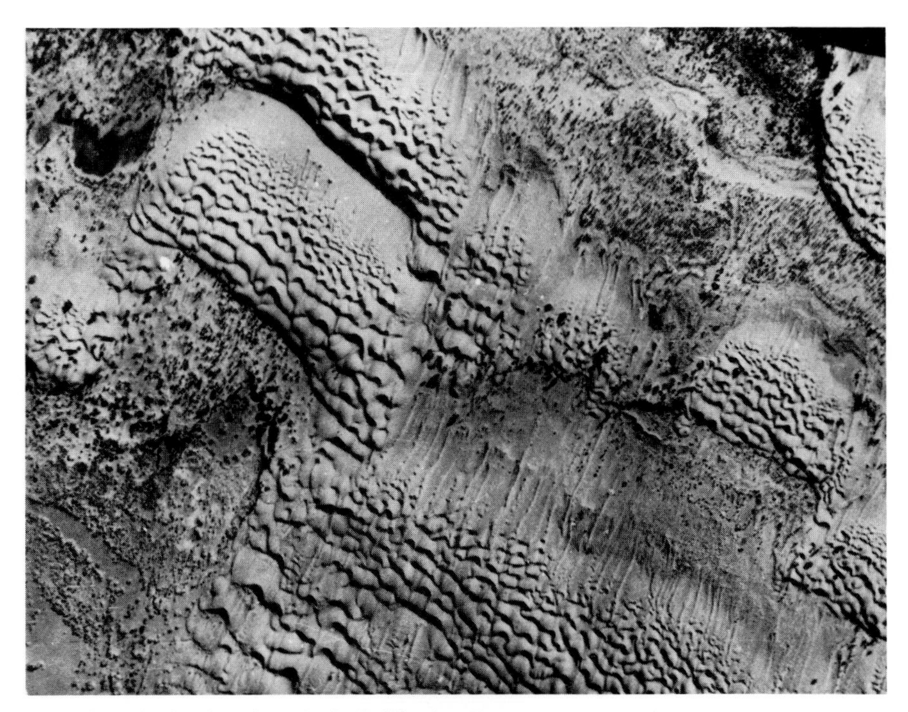

Fig. 1. Complex barchan dunes in the Taklimakan Desert.

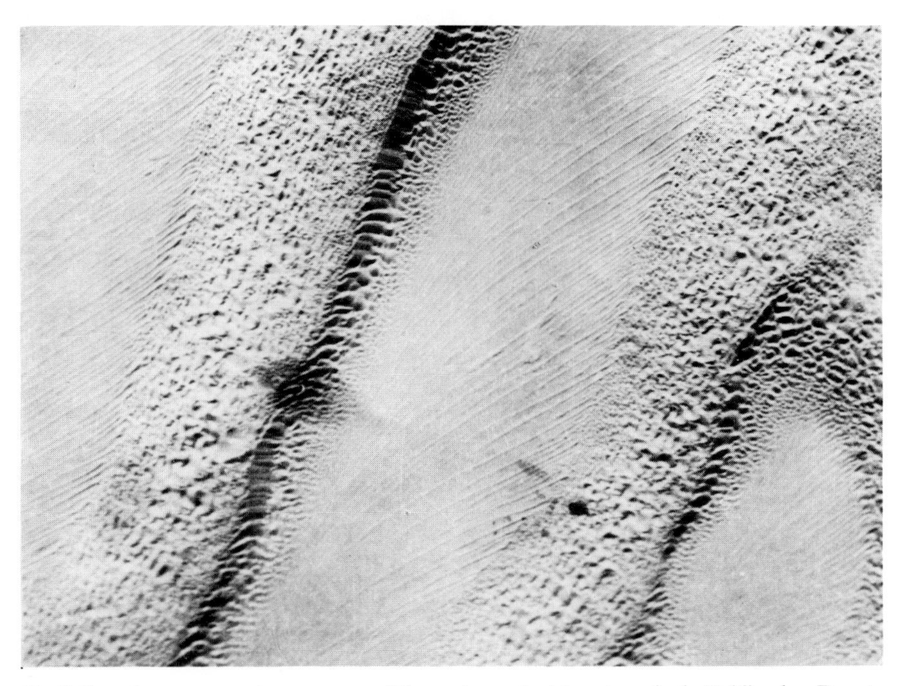

Fig. 2. Complex transverse dunes are one of the most prevalent dune types in the Taklimakan Desert.

Fig. 3. Complex longitudinal dunes in the Taklimakan Desert.

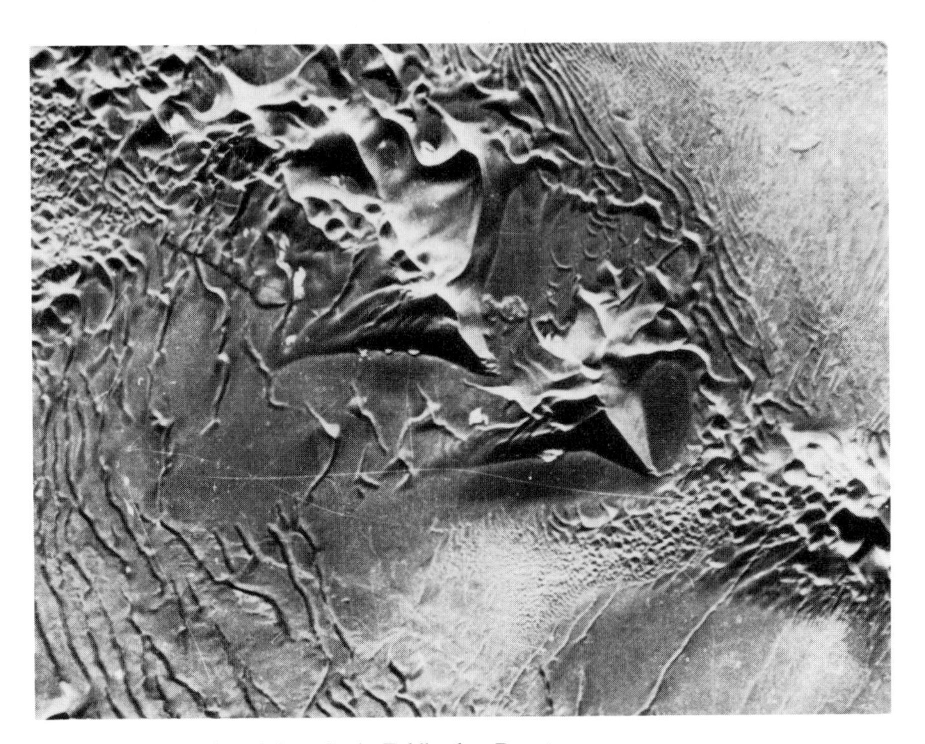

Fig. 4. Horn (cone) shaped dunes in the Taklimakan Desert.

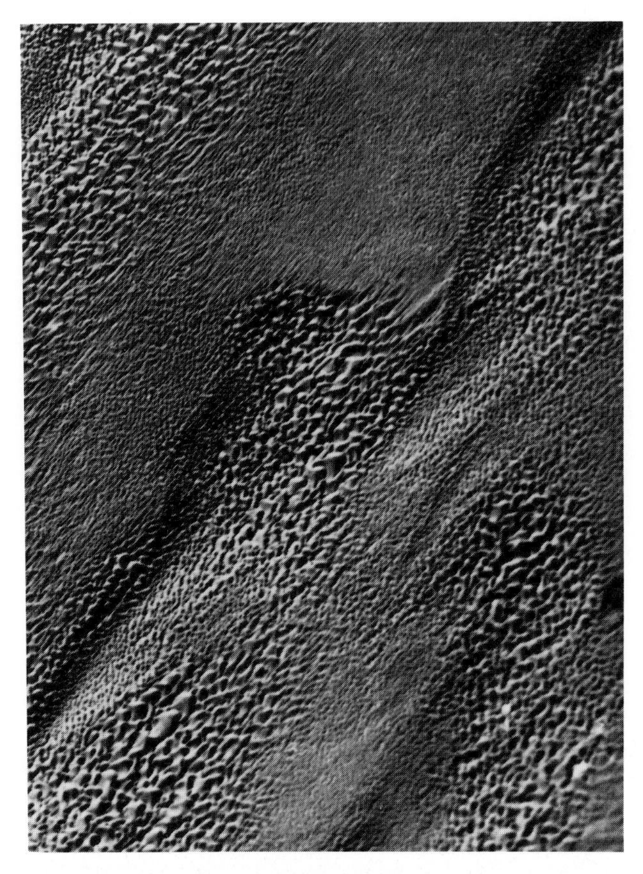

Fig. 5. Scaled dune groupings resemble the scales of fish when viewed from above. They are similar to longitudinal dunes because they parallel the prevailing wind.

the prevailing wind.

6) Dome shaped dunes (Fig. 6) primary characteristics are: symmetrical slopes, with secondary dunes imbricated upon them. The slipslope is without any conspicuous curves, and its length and width are nearly equal. Their height ranges from 40–60 m. On a planar surface the shape is usually circular or elliptical. In general, they are irregularly distributed but is some parts of the desert this type of dune may be linked together with others of this type while retaining its dome shape.

Landforms in the Taklimakan Desert are often the result of combinations of barchanic ridges, but there are also different complex types of sand dunes which are described above.

Although the Taklimakan Desert is characterized by the above-mentioned types they can be devided into three main types according to the relationship between the morphology and the wind.

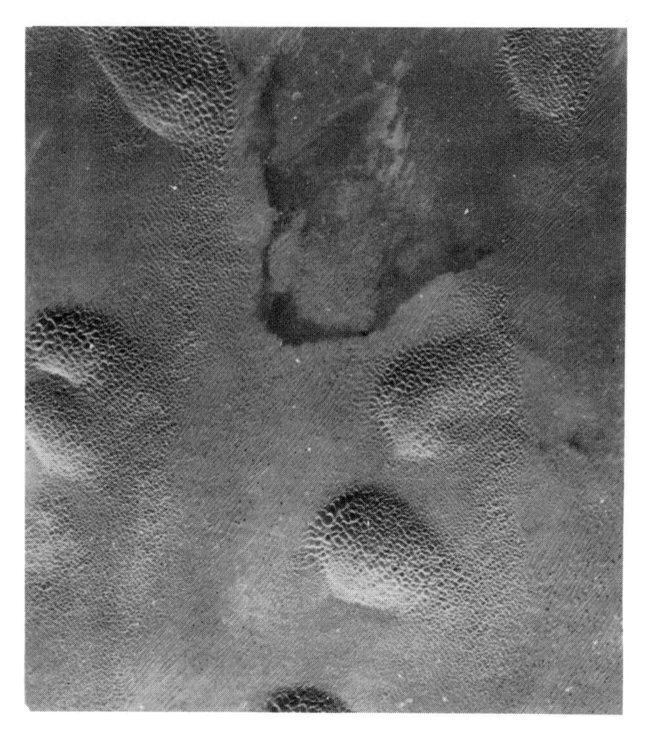

Fig. 6. Dome shaped dunes with smaller dunes superimposed on them in the Taklimakan Desert.

Transverse dunes

The morphology of the dune is directed such that it is vertical with the sand-blown wind so that it forms a crisscross angle of 60 to 90 degrees. Some examples of this are the compound barchanic dunes, compound dune chains and barchanic chains.

Longitudinal dunes

The morphology of these dunes is oriented parallel with the direction of the sandblown wind, namely, it forms a crisscross angle of 30 degrees. Examples of this type are compound longitudinal dunes, longitudinal dunes and barchanic longitudinal dunes.

Dunes under the effects of a multi-directional wind

The morphology of dunes of this type is not directed parallel with, or vertical to any wind direction but is formed generally by uni-directional winds. In this case the wind power is similar, and it is characterized by crest lines and slipslopes of different directions. This type of dune is typified by pyramid dunes (star dunes).

The movement of the sand dunes

According to observation, the sand dunes in the Taklimakan Desert can only be moved by a wind velocity greater than five meters per second. In general, this effective wind (sand blowing wind) comprises only one third of the annual wind velocity. The movement of the sand dunes depends on the nature and the strength of this kind of wind.

Observations show that the direction in which the sand dunes move conforms to the direction of the 'sand-blowing wind'. The general direction of the movement is similar to that of the annual resultant of wind direction. In the southwestern part of the desert, from south to north, the movement of the sand dunes changes its direction from ESE, to SE and then SSE. In the eastern part of the desert, the direction changes from W to WSW and then SW. In the central part of the desert, the direction changes to SSW, and in the northwestern part of the desert, it changes from SW to SSW. The manner of the movement differs according to the complicated nature of the direction of the sand blowing wind. In the eastern and western parts of the desert where a uni-directional wind prevails, sand dune movement is of the progressive type. In the sourh and the nortwest, the movement is of the oscillatory progressive type.

The regime of the sand blowing wind not only determines the direction of movement of the sand dunes, but it also has an influence on the rate of movement of the sand dunes.

Firstly, the rate of movement of the sand dunes is related to the wind velocity. The movement of the sand dunes takes place through the movement of the sand grains. When the wind velocity exceeds threshold velocity, the quantity of sand movement in a given time unit increases with the increase of wind velocity. Because the wind velocity is closely related to the rate of movement of sand dunes, the movement of the dunes occurs predominantly during the windy seasons of spring and early summer. Data from observations in the southwestern part of the desert show that 60 to 80% of the annual movement of the sand dunes took place during the windy seasons.

Secondly, the direction of the wind and its regime exercise some influence on the rate of the movement of the dunes. Observations show: under a uni-directional wind, the rate of movement of the sand dunes is faster than under a multidirection wind. This is due to the fact that any wind with a given strength and direction will create a corresponding profile and horizontal form. When the wind changes so that it no longer corresponds to that indicated by the original form of the sand dunes, it remodels the original form so that it corresponds to the new direction and force of the wind. Therefore, in the region where a multi-directional wind exists, when the wind changes direction, the energy of the wind is first used for the remodelling of a new form. The advance of the sand dunes does not require a great deal of wind force. This is the reason that the rate of advance of sand dunes is relatively slow.

Although the movement of sand is controlled chiefly by the wind, the rate of the movement of the sand dunes varies. This variation is caused by the following factors:

1) The influence exerted by height of the sand dune on the rate of movement. Observations show in the western part of the desert, that with different ground conditions, and different densities of sand dunes, the relationship between the rate of the movement of a sand dune and its height is inverse. The coefficient correlation is high, generally over 0.8.

2) Water condition. Under humid conditions, the viscosity and cohesion of sand grains is higher, so a higher threshold velocity is required. Observations in the southwest of the desert show that after rain, when the water content on the surface of the sand dunes (grain diameter 0.25–0.1 mm) is greater than 4‰, the sand will not be moved if the average wind velocity is less than 11 meters per second. Only after the wet sand has been dried by strong wind can the sand begin to move. It is evident that the presence of water exerts an influence on the rate of movement of the sand dunes.

3) The condition of the vegetation coverage. The presence of a vegetative covering on the sand dunes tends to diminish the rate of movement of the sand dune or to stop its movement altogether. This is due to the fact that vegetation increases the roughness of the land surface, weakening wind velocity near the ground and diminishing the transport of sand.

4) The influence of the size of sand grains on the rate of the movement of the sand dune. When the grain diameter is 0.1–0.25 mm, the threshold velocity is 4 meters per second (at 2 meters height). When the grain diameter is 0.25–0.5 mm, the threshold velocity is 5.6 meters per second. When the diameter is 0.5–1.0 mm, the threshold velocity is 6.7 meters per second. A sand dune composed of fine grained sand will move faster than one of coarser grain.

Because the rate of movement of the sand dunes is influenced by the above factors, and these factors are not the same everywhere in the Taklimakan Desert, the rate of dunal movement is very complicated. Based on observations in the desert, studies from air photos and land survey, the rate of the movement of the dunes may be classified into the following types:

1) Slow rate type: with a migration rate of 1 m per year; a good example of this

type is the big complex sand dunes in the inner part of the desert.

2) Middle rate type: with a annual rate of migration of 1–5 m, examples can be found in the western, northern and southeastern parts of the desert, and also at the two banks of the river reaching far into the desert.

3) Relatively great rate type: the annual rate of migration is 5–10 m with the best examples located in the eastern part of the desert.

4) Great rate type: the annual rate of migration is 10 m, with the best examples visible in the marginal zone of the southwestern and southeastern parts of the desert.

Author's address:
Lanchow Institute of Desert Research
Academia Sinica
Lanchow, China

8. North American deserts

Harold E. Dregne

The deserts of North America are distributed over northern Mexico and the western United States between about 23° and 45° north latitude [Fig. 1]. They owe their existence to a combination of causes. The orographic [rain-shadow] effect is primarily responsible for the aridity of the Chihuahuan Desert. In the Great Basin, there is a winter maximum in rainfall. In the Chihuahuan Desert and the Sonoran Desert, the rainfall maximum comes in summer.

As the term is used here, the 'desert' region delineated on the map of Figure 1 includes the hyper-arid and arid climatic zones and the drier part of the semiarid zone shown on the UNESCO map of the world distribution of arid regions [1]. The boundary line approximates the division between grazing lands and potentially rain-fed cultivated lands. The Chihuahuan Desert and the Sonoran Desert are vegetative regions, not climatic regions. The Great Basin, on the other hand, is a geographic unit lying between the Rocky Mountains on the east and the Sierra Nevada and Cascade Mountains on the west. Within the Great Basin is Death Valley which, along with an area north and south of Yuma, Arizona, constitutes the only hyper-arid climatic zone (the true climatic desert) in North America. Nearly all of the desert lands of Figure 1 belong to the intermountain phytogeomorphic region.

Vegetation

Desert regions in North America have two general types of vegetation, one in the cool northern regions and the other in the hot southern regions. Sagebrush, (*Artemesia tridentata*) and saltbush (*Atriplex confertifolia*) dominate the cool region plant communities of the Great Basin whereas creosotebush (*Larrea tridentata*) dominates the large communities in the hot regions of the Chihuahuan and Sonoran deserts. The southern communities also have mixed stands of large shrubs, small trees and a variety of succulents. An understory of grasses is present under natural conditions, in both the cool and the hot climates, but is very sparse in the hyper-arid zones. The diversity of flora is greatest in the

El-Baz, F. (ed.), Deserts and arid lands. ISBN: 90-247-2850-9.

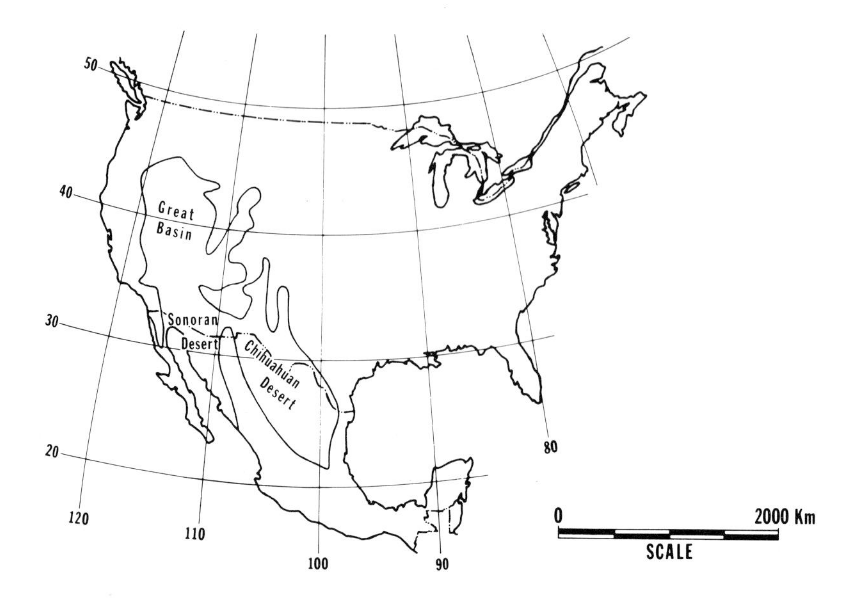

Fig. 1. Deserts of North America.

Sonoran Desert, intermediate in the Chihuahuan Desert, and least in the Great Basin [1].

Soils

Soils consist mainly of Aridisols and Entisols, with inclusions of Vertisols in the closed depressions (playas) and of Mollisols on the cooler and wetter fringes of the desert zone. Medium- to coarse-textured gravelly and stony soils are common, particularly on either side of the Mexico-United States border. Soils are alkaline, calcareous, low in organic matter, weakly developed and nonsaline, for the most part.

Climate

Moisture distribution patterns in the arid regions of the United States and Mexico are heavily influenced by the presence of the Pacific Ocean on the west and the Gulf of Mexico on the east. Summer moisture comes from the Gulf, winter moisture from the Pacific. Within the desert region of Figure 1, there are no moisture patterns that are of the pronounced Mediterranean wet winter-dry summer or the Sahelian wet summer-dry winter type. A rather unusual pattern of a high summer peak, a low peak during winter, and a dry spring is found from south central New Mexico to the Gulf of California. All months of the year usually are moisture deficient in that potential evapotranspiration exceeds precipitation throughout the desert zone.

Table 1 gives climatic data for six locations in the desert zone. The stations are arranged approximately in a north-south order. Guaymas, even though it is on the Pacific coast of Mexico and on the west side of Sierra Madre Occidental, has a summer-maximum rainfall pattern, as does Chihuahua on the east side of the Sierra Madre Occidental. As is apparent from Table 1, seasonal differences in rainfall are considerable among the various stations, as are annual differences (Table 2). Rainfall variability from year to year is greatest where the average annual precipitation is low and least where the annual average is highest (Table 2), whereas annual temperature variations are small (Table 3). Freezing temperatures can be expected to occur in some years everywhere except close to the Gulf of California. The large rainfall variations between and within years that is typical of the dry climates places severe adaptive stresses on the native plants and animals and causes wide swings in agricultural productivity from year to year in the better endowed climatic zones. In the hyper-arid zones, native productivity is always low.

Impact of man

Man has brought about a major alteration in the plant, animal and soil components of the desert landscape of North America in the past 100 to 200 years. Most of the alteration has changed the natural environment while producing wealth and making living conditions easier for the human occupants. Many of the changes are, for all practical purposes, irreversible. The giant holes in the ground at open-pit copper mines, exhausted ground water aquifers, the transportation routes, and soils altered by accelerated erosion and salinization have an indelible effect on the plant and animal environment. For the ecologically-minded, the changes represent degradation; for developers they may seem to represent improvement upon nature. The man-induced soil and vegetation modifications described in this chapter have lowered the native productivity of the land for food and fiber and have increased pollution of the air, land, and water. Overgrazing, woodcutting, improper land and water management, mining and recreation are the land degradation (desertification) processes to be discussed [3, 4].

Table 1. Climatic data, selected stations.

Station	Elevation (m)	Temperature extremes (°C) Max.	Min.	Precipitation (mm) J	F	M	A	M	J	J	A	S	O	N	D	Total
Salt Lake City, Utah	1286	42	−34	36	31	41	56	38	33	18	23	18	31	33	36	394
Las Vegas, Nevada	659	47	−13	13	8	8	8	3	3	10	13	8	5	10	10	99
Tucson, Arizona	728	44	−14	20	18	15	10	3	5	56	51	35	18	15	23	269
El Paso, Texas	1194	43	−22	10	10	10	8	5	15	38	28	31	20	8	13	196
Fort Stockton, Texas	900	46	−22	15	15	10	20	41	38	36	33	36	33	15	13	305
Chihuahua, Chih.	1350	39	−11	5	10	8	5	5	43	91	94	84	23	13	10	391
Guaymas, Sonora	18	47	5	10	5	5	3	tr	3	43	69	53	18	8	20	237

Table 2. Precipitation variability at four U.S. stations.

	Annual precipitation (mm)			Ratio:
	Average	Wettest year	Driest year	wettest/driest
Las Vegas, Nevada	99	271	14	19.4
El Paso, Texas	196	444	61	7.3
Tucson, Arizona	269	531	120	4.4
Lubbock, Texas	459	1030	222	4.6

Table 3. Temperature variability at four U.S. stations.

	Annual temperature (°C)			Ratio:
	Average	Warmest	Coolest	Warmest/coolest
Las Vegas, Nevada	18.4	19.6	14.7	1.3
El Paso, Texas	18.0	19.5	17.1	1.1
Tucson, Arizona	19.7	21.4	18.9	1.1
Lubbock, Texas	15.3	17.3	14.4	1.2

Desertification

The term 'desertification' is currently used for the processes of land degradation due to man's activities that can – if carried to the extreme – change productive land into a wasteland. The definition of desertification used here is the following:

> Desertification is the process of impoverishment of terrestrial ecosystems under the impact of man. It is the process of deterioration in these ecosystems that can be measured by reduced productivity of desirable plants, undesirable alterations in the biomass and the diversity of micro and macro fauna and flora, accelerated soil deterioration, and increased hazards for human occupancy [5].

Much disagreement exists concerning what desertification means. Many people think of desertified land as land that is completely useless, either due to man's activities or to drought [6]. That is not the meaning used here or at the 1977 United Nations Conference on Desertification. Rather, desertification is a process that has a number of stages, from slight to very severe. Only the latter stage resembles a man-made wasteland. Fortunately, such lands are inextensive, at present. Slightly desertified land can usually be restored to full productivity rather easily; very desertified land may be so badly degraded that restoration is not economically feasible.

The extent and magnitude of desertification in North American deserts is

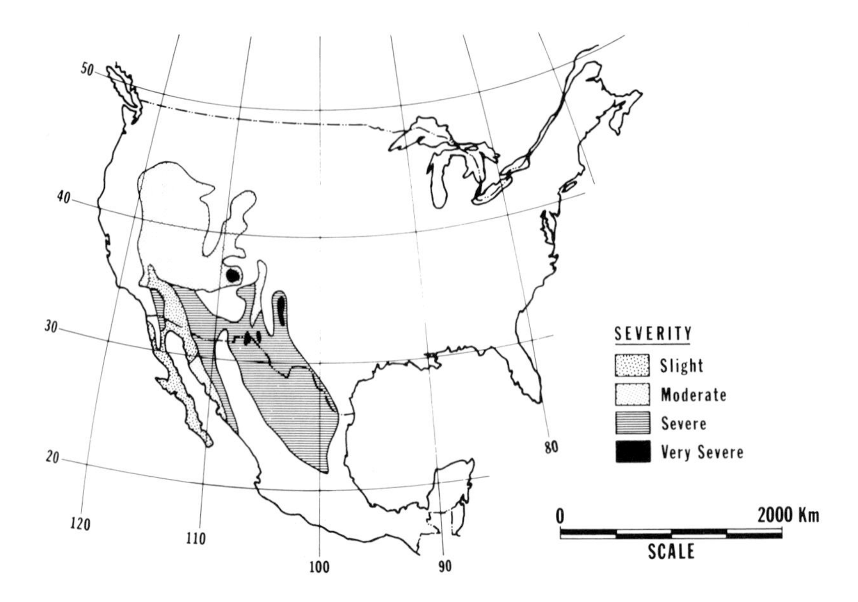

Fig. 2. Desertification in North American desert zones.

shown in Figure 2. It may appear paradoxical that the hyper-arid zones (the climatic deserts) of Death Valley and around Yuma are mapped as only slightly desertified, but that stems from the definition of desertification as man-made degradation. The hyper-arid regions are natural deserts, not the results of man's activities. Except for small areas degraded by mining, recreation, urban, and irrigation developments the region is too inhospitable to have been affected by man. Only the salinized Imperial and Mexicali valleys are large enough to be shown on the small scale map of Figure 2.

Overgrazing

By far the most extensive form of desertification is that caused by overgrazing by livestock. Settlement of North American deserts began in Mexico in the 1700s, then moved northward into what is now the United States in the early 1800s, meeting settlers from the eastern United States. By the middle of the 19th century, heavy stocking with cattle and sheep had already begun to degrade the ranges. Within a few decades, land degradation – in the form of an increase in undesirable shrubs and an associated decrease in grass – had become a serious problem near watering points where livestock were concentrated. As overgrazing continued, much of the remaining grassland deteriorated. By the beginning of the 20th century, desertification had left its mark on virtually all of the arid rangelands. Deterioration of the vegetative cover was followed by the acceleration of wind and water erosion and the appearance of man-made gullies and sand dunes. Nearly 85% of the rangelands of North America exhibit moderate or worse desertification.

Deterioration of the vegetative cover in the cool arid regions of the northern United States and central Mexico seldom was as severe as it was in the hot arid regions of the southern United States and northern Mexico. Increases in shrub density and shrub invasion of grasslands due to overgrazing had a marked impact in the southwestern United States and the neighboring states of Mexico. Wind and water erosion in the deteriorated rangelands produced changes in soil aggregation and permeability that adversely affected the soil moisture status and made the microenvironment less favorable for grasses. The soil environment became more xeric, even though the annual rainfall had not changed, due to the greater erosion and runoff on degraded lands. York and Dick-Peddie [7] postulated that most of the southern part of the state of New Mexico has been changed from a desert grassland ecosystem (less xeric) to a desert shrub ecosystem (more xeric) by soil deterioration during the past one hundred years. The phenomenon is the same as has been observed in the Middle East, where more xeric soil conditions in the hills of Lebanon, Israel and Jordan have led observers to believe that the climate has become drier during the past 2000 years. Excessive erosion has made the soils less able to absorb rain, and the result is increased runoff and less moisture in the soil for plant growth. The soil climate has, in fact, changed, not because of any change in atmospheric conditions. In one experiment studying the effect of grazing on water intake and runoff, the water intake rate on the lightly grazed land was nearly three times greater than on the heavily grazed land and runoff was nearly twice as high under heavy grazing conditions [8]. That phenomenon has been observed in grazing lands throughout the world.

Four grazing areas in Figure 2 have experienced very severe desertification that has essentially ruined the land for the foreseeable future. All lie in the southwestern United States and one extends across the border into Mexico. Three of the areas were productive grasslands before they became infested with mesquite (*Prosopis* spp.) and subjected to excessive wind erosion. Now the land is covered with hummocks and small dunes held in place by mesquite bushes; what little grass there is can be found only under the bushes. The nothernmost very severely desertified area is in the Navajo Indian Reservation in New Mexico and Arizona. There, on soils derived from shale, overgrazing led to severe water erosion, loss of the thin topsoil, and developments of large arroyos (gullies). It is too late now for controlled grazing to restore the land to its former productivity, which was never high.

Rangelands of the United States are in mostly fair to poor condition (Table 4), and it is likely that the same situation prevails in Mexico. The very slow progress being made in improving them indicates that the task is not easy when economic benefits are on a low per-hectare basis.

Table 4. Range conditions of federal lands.

Year	Range condition class (in %) Good or excellent	Fair	Poor or bad
1936	16	26	58
1966	18	49	33
1972	18	50	32

Source: Box [16]

Woodcutting

In and around the desert region of North America, wood has been cut predominantly for two purposes: fuel and mine timbers. Mining operations which have been extensive in the arid regions of northern Mexico and southwestern United States, required timber to support tunnels and firewood for smelting ore. Two conifers, the pinyon (*Pinus edulis*) and juniper (*Juniperus* spp.), probably have been used more in mining operations than any other zerophytic trees. Numerous shrubs have provided firewood. However, woodcutting in North America has not had as destructive an effect on the environment as in most arid regions, because oil, kerosene, and coal largely supplanted wood as fuel many decades ago.

Salinization

Salinization and waterlogging are the twin evils of irrigated agriculture. Irrigation of arid regions has been practiced for centuries in North America [9], but there is no indication that salinization and waterlogging ever were in the past the threats that they were in Mesopotamia [10]. Trouble came to the forefront in the last 50 to 100 years when big dams and large reservoirs provided an assured – and copious – supply of water all year. Today, about 30% of the irrigated land in Mexico and 10% of the United States is affected to some degree by salinization, most of it associated with waterlogging. In the desert zones, the percentages would be considerably higher. Salinization is worst in the upper reaches of the Colorado River, in the Mexicali Valley, and along the Pecos River in Texas, but no irrigated region is free of the problem. Without a doubt, salinity is the greatest threat to the long time survival of irrigated agriculture.

When irrigation was begun by Europeans in North America, little thought was given to the character of the land or to the need for drainage, and the settlers – except for a few from Spain – had no previous experience with irrigation. As a consequence of the lack of knowledge of the hazards posed by irrigation, salty and poorly drained soils were irrigated simply because they occupied low-lying level areas that were easy to water. Even in later years, when governments developed large irrigation projects, few engineers anticipated the onset of the drainage problem that came about when water tables rose following the application of excessive amounts of irrigation water. Irrigators everywhere apparently share a similar philosophy: if a little water is good, more is better. Inevitably that leads to overirrigation as long as the water supply lasts and water costs are low.

Important though salinization has become, abandonment of irrigated land because of salt has not been widespread. In the two countries, probably less than 5% has been allowed to lie idle. Of more economic significance is the crop yield reductions brought about by slight to moderate soil salinization [11]. Effective techniques have been developed to cope with salinity problems in all but the most impermeable soils. A more pervasive problem is the increase in salinity of the irrigation water in its passage down the river [12]. Every time water is diverted from the Rio Grande (Rio Bravo), for example, to meet industrial, domestic, or agricultural needs, the portion that returns to the river is more saline than it was before. This means that water users located at the lower end of a river must contend with more salinity than users at the upper end. That fact of life had endangered disputes between the United States and Mexico on the quality of the water diverted from the Colorado River for use in the Mexicali Valley. The dispute has been resolved amicably – though expensively – but the salinity of water delivered to the states of Arizona, California, and Baja California will rise inexorably in the next 20 years as water development continues in the upper basin states and return flows become more saline. The battle to control salinization and waterlogging is never-ending. As with overgrazing, salinization of irrigated land is a greater problem in the hot deserts than in the cold deserts, due to the higher rates of evapotranspiration in the hot deserts.

Erosion

Accelerated water erosion accompanied the overgrazing that attained serious proportions a hundred or more years ago. Mining operations also led to increased erosion but the affected area was much smaller than that caused by excessive livestock concentrations. An unknown amount of desert pavement formation followed accelerated sheet erosion on gravelly soils, and increased arroyo (gully) formation occurred at about the same time [13]. Desert pavements help to stabilize soils against further water or wind erosion and are, therefore, desirable on overgrazed lands or in the hyper-arid regions where vegetative cover is naturally sparse.

Wind erosion became worse as overgrazing intensified but water erosion has always been – and remains – more serious in the desert zone.

A relatively new development in desertification in the United States is the growing use of off-road vehicles for recreational purposes [14]. Along with the increase has come land abuse, particularly in the California desert east of Los Angeles. Most of the abuse has occurred on state and federal lands near population centers. Drivers of off-road vehicles believe that they should be allowed to use public desert lands for their own enjoyment. Many of them undoubtedly think that deserts are wastelands of little or no value and that there is no reason to be concerned about their abuse. The destructive philosophy reminds one of early settlers in the Americas who were unconcerned about land degradation because they could always move on to new land.

Regulations to control land abuse by off-road vehicles on public desert lands have been issued but obtaining compliance is very difficult. As a consequence, there is growing sentiment for setting aside 'sacrifice' areas where recreational abuse would be allowed, while protecting the remaining land. The very concept of a sacrifice area is repugnant to many environmentally conscious people. There is evidence that areas of intensive off-road vehicle use in the California Desert have become sources of immense dust storms, observable on satellite images.

Encroachment of wind-eroded soil on settlements is only a minor problem in North American deserts, but air pollution and sand accumulations on highways are important. Highways in Arizona, New Mexico, and Texas must be closed to traffic from time to time in order to limit automobile accidents during dust storms. Removal of sand from highways is another cost associated with wind erosion.

Conclusions

Desert zones of North America, from the state of Oregon in the north to the state of Guanajuato in the south, possess a rich variety of plants, animals, soils, topography, geological materials, and climates. Much of the mineral wealth of Mexico and the United States has come from the desert regions, along with cattle, sheep, fruits, and vegetables. They lie in the sunbelt where cities are growing rapidly as people seek new opportunities and pleasant climates in which to live.

Man has had a profound effect on the natural environment of the deserts, to the point where the modern landscape nearly everywhere has a different appearance than it did as little as a hundred years ago. Vegetation differences are the easiest to detect, although the changes would be observable only to the trained eye in and around the hyper-arid zones and in the northern part of the cool deserts. In the brief span of time since European occupation made its imprint on the desert, some lands have already gone through a cycle of irrigation and abandonment, plant and animal species have been brought to the edge of extinction, whole ecosystems have become more xeric, tens of thousands of hectares of land have been covered with asphalt and buildings, water supplies have been irrevocably depleted, and the standard of living for most inhabitants has improved while the quality of the environment has deteriorated.

Verification of the magnitude of the desertification that has occurred in these deserts zones is very difficult except for the case of localized effects due to urbanization, mining, transportation networks, and irrigation. Good records go back only a few decades, which has made it necessary to rely upon published accounts of how conditions were a century or two ago. That is highly unsatisfactory. Photographic records of the kind used to compare vegetative conditions over a 30- to 70-year period in the Arizona Desert [15] are rarely compiled because of the labor involved in doing so. Photographs, however, are an excellent device for making comparisons over the past 100 years or so.

Monitoring future extensive changes in vegetative cover, water and wind erosion, and salinization and waterlogging is likely to require a long time in the desert zones. A degree of stability probably has been established in the grazing lands and in the irrigated lands, with most of the land damage already done. This is not true in smaller areas where recreation, mining and urbanization are still affecting the land, air, and water resources. Those effects can be felt in a matter of months or even weeks, whereas changes in the regional vegetation are confounded by drought events and, therefore, must be monitored for many years in order to ascertain whether real changes due to management have occurred.

References

1. Unesco: Map of the world distribution of arid regions. MAB technical notes no. 1 (1:25,000,000 map). Paris: Unesco, 1979, 54 pp.
2. McGinnies WG: Appraisal of research on vegetation of desert environments. In: McGinnies WG, Goldman BJ, Paylore P (eds) Tucson: University of Arizona, 1968, pp 381–566.
3. Médellin-Leal F (ed): La desertificación en Mexico. San Luis Potosi, Mexico: Instituto de Investigación de Zonas Deserticas, Universidad Autónoma de San Luis Potosí, 1978, 130 pp.
4. Sheridan D: Desertification of the United States. Washington, DC: Council on Environmental Quality, 1981, 142 pp.
5. Dregne HE: Desertification: man's abuse of the land. Journal of Soil and Water Conservation (33):11–14, 1978.
6. Glantz MH (ed): Desertification. Boulder, Colorado: Westview Press, 1977, 346 pp.
7. York JC, Dick-Peddie WA: Vegetation changes in southern New Mexico during the past 100 years. In: McGinnies WG, Goldman BJ (eds) Arid lands in perspective, Tucson: University of Arizona Press, 1969, pp 156–166.
8. Rauzi F, Hanson CL: Water intake and runoff as affected by intensity of grazing. Journal of Range Management (19):351–356, 1966.

9. Donkin RA: Agricultural terracing in the Aboriginal new world, Viking fund publications in anthropology 56, 1979, 196 pp.
10. Jacobsen T, Adams RM: Salt and silt in ancient Mesopotamian agriculture. Science (128): 1251–1258, 1958.
11. Aceves Navarro E: El ensalitramiento de los suelos bajo riego. Chapingo, Mexico: Colegio de Postgraduados, 1979.
12. Dregne HE (ed): Managing saline water for irrigation. Lubbock, Texas: International Center for Arid and Semi-Arid Land Studies, Texas Tech. University, 1977.
13. Cooke RV, Reeves RW: Arroyos and environmental change in the American southwest. Oxford, England: Clarendon Press, 1976.
14. Baldwin MF, Stoddard DHJ: The off-road vehicle and environmental quality: Washington DC: The Conservation Foundation (2nd edition), 1973.
15. Hastings JR, Turner RM: The changing mile. Tucson: University of Arizona Press, 1965.
16. Box TW: The arid land revisited – One hundred years after John Wesley Powell, Logan: Utah State University, Faculty Association, 1977.

Author's address:
International Center for Arid and Semi-Arid Land Studies
Texas Tech University
P.O. Box 4620
Lubbock, TX 79409, USA

9. Natural resource survey and environmental monitoring in arid-Rajasthan using remote sensing

H.S. Mann, K.A. Shankaranarayan and R.P. Dhir

Abstract

Characterization and assessment of natural resources for development planning and technology transfer have been one of the major activities in the Indian arid zone and, as of now, over 87,000 km^2 of arid Rajasthan have been surveyed in this way. All along, aerial photointerpretation has played a key role by greatly enhancing the speed and accuracy of this endeavor. Whereas landforms, soils and land use, and to some extent vegetation are easily and directly discernible from the photographs, mapping of other resources is facilitated mainly by the relationship that they have with the land attributes appreciable from the imagery. Experience has shown that groundwater exploration efforts can be made more purposeful by restricting detailed investigations to areas indicated as promising by photoanalysis. The usefulness of multispectral data from Landsat in source survey and monitoring have been demonstrated. Particular advantage from repetitive coverage afforded by satellite is helpful in estimating biomass and planning rational utilization.

Unlike many arid zones, Rajasthan has high human and livestock pressures and under both, the major land uses, namely arable farming and open grazing, the present management is somewhat exploitative of the natural assets. However, according to present indications, the rate of deterioration fortunately is slow. Therefore, monitoring this change while faced with a large interannual variation in various desertification manifestations, is nothing short of a challenge. The present approach requires conjunctive use of closely spaced ground observations in selected sample sites, repetitive low altitude photography, and satellite sensing for large area applications.

Introduction

The arid tract in Rajasthan covers an area of 19.6 million hectares, which is 62% of the total hot arid zones in the country. It has a population of 13 million people

El-Baz, F. (ed.), Deserts and arid lands. ISBN: 90-247-2850-9.

with an average density of over 60 persons for every square kilometer area. This makes Rajasthan one of the most thickly populated tracts of all the regions of comparable aridity in the world. Besides supporting a large population, the arid tract makes many notable contributions to the national economy. Nearly forty percent of the wool produced in the country comes from this tract. Cattle, sheep, other animal products and minerals are items worth mentioning in this regard. Due to these considerations, the region has been the site of various programs for the socioeconomic development of the country on scientific lines. As a result, an impressive infrastructure exists from research and development efforts in this region, and physical achievements are evident. Right from the beginning of this effort, appropriate emphasis was placed on characterization and assessment of natural and human resources for developmental planning and the transfer of technology. As a result, a wealth of information and equipment has been built up over the years in remote sensing, particularly in aerial photography, which has played a key role in this activity. Though concern for deteriorating ecological conditions in arid zones has been felt all along, the subject has come under sharper scrutiny in the wake of the recent worldwide attention to this problem. Here again, an important role is envisioned for remote sensing technology. This paper will discuss these experience in the Indian arid zone.

Setting of the area

The tract receives a mean annual rainfall that ranges from 100 mm in the extreme west, to 450 mm in the east, where the potential evaporation rages from 1800 to 1600 mm. Of the total rainfall, 83 to 95 percent is received during the monsoon period from July to September. Though the rainy season sets in with incredible precision, the actual amount of rainfall received is open to variation from year to year, the coefficient of variation being 40% in the east and 70% in the western parts. The region has a strong wind regime, the mean wind speed in June, the windiest month, is 14 to 27 km per hour.

Most spectacular amongst the landforms are the dunes, which cover 58% of the region in varying degrees of frequency. Associated with the dunes are large stretches of light textured sandy plains dominated by Torripsamments and coarse loamy Camborthids. There is a sizeable area of Paleorthids or hard-pan soils. The central and sourthern part is made up of medium and fine textured Camborthids and Calciorthids.

Ecologically, the vegetation of the region can be described as tropical thorn forest and xeromorphic woodland types with many subdivisions based on variations in habitat. However, this natural vegetation cover has been greatly transformed under prolonged biotic influence. With the exception of the westernmost districts, 69 to 90% of the arid zone is under cultivation though in a system of fallow farming. In the more arid tract, open grazing is the dominant landuse. However, agriculture in the region is essentially a mixed farming enterprise with most households maintaining a sizeable herd of livestock that is sustained on crop residues, grazing on fallow lands and topfeed species that are consciously maintained in crop lands.

Arising from the paucity of rainfall, surface and groundwater resources are scarce. In few favorable locations important aquifers have been located, but in most of the major areas, groundwaters are deep-seated and also of poor quality. Presently, a major project based on diversion of waters from the Indus Basin is nearing completion, which will command 1.14 million hectares of the most desertic tract in the region.

Aerial photography in natural resources survey

The need for an inventory of natural resources as an aid to developmental planning was realized at the very beginning of the scientific effort in the region. It was further felt that the best results could accrue only from a multidisciplinary survey covering landforms, soils, vegetation, surface and groundwater and present landuse. Since human, livestock and wildlife resources are also important, the survey covers these aspects as well. In this activity remote sensing, notable analysis and interpretation of aerial photographs and recently also Landsat outputs, have played a key role. These tools have been of invaluable help in improving the accuracy as well as the speed of the surveys by greatly cutting down the field work, which is otherwise very strenuous and time consuming because of the poor accessability of the terrain and a lack of camping sites.

Use in landform analysis

Since landform features are the most conspicuous, aerial photos have been used with great results in geomorphological studies. With experience it has become possible to identify and delineate various landform units like the aggraded alluvial plains, bare and buried pediments, piedmont plains, hills, plateaux, playas, and such others [1–4]. Particular mention should be made here of the use made in recognizing various dune types and their distribution, orientation, size, etc. [5, 6]. Wherever a drainage network exists, aerial photos have offered a unique means of quantifying it in terms of drainage density, order of streams, bifurcation ratio, individual stream length, etc. [7, 8]. The information is applied in the assessment of surface water resources.

Use in mapping soil resources

Tone, texture and pattern on aerial photos have been successfully used in reconnaissance soil mapping. The approach used has been to analyze the photos based on the above characteristics and then to ground check individual analytical units for characterizing the soils [4]. Dhir [10] found that in addition to these, the associated features like distinctness of field boundaries, degree of mottling due to scrub vegetation, density of tree vegetation, landuse manifestation, could also be advantageously used in the delineation of soil boundaries. In some situations depth classes of soils over a calcareous pan can be discerned but in other cases extensive field work is unavoidable.

Many works have shown the ease with which salt-affected soils can be mapped from aerial photos. Further, Dhir and Singh [11] and Kolarkar et al. [12] have used the distribution pattern as seen from the photos in inferring the categories of salt-affected soils and possible factors responsible for the formation of these soils.

Use in hydrogeological studies

The extent to which geological structures and fabric features of the various lithological formations can be seen when viewed from the aerial photos, directly determines its value in assessing the groundwater potential. Experience in arid zones [13, 14] shows that of still greater advantage is the relationship that exists between a groundwater situation and geomorphic features. This is because to a great extent the factors responsible for the development of landforms are also the factors contributing to the development of water bodies and their quality. Many important parameters like depth to static water level, total soluble salt content and the effect of structurally weak zones on water potential were seen as being significantly influenced by the geomorphological setting such as the landform type, nature and the thickness of a surface formation and the surface drainage [15]. Since many of these geomorphic features are easily identifiable from aerial photos, the role of the latter in groundwater becomes obvious. However, this does not rule out the use of electrical and seismic methods in groundwater prospecting; what is suggested is that with proper aerial photointerpretation, subsequent investigations can be made far more effective for the time and money spent.

Use in vegetation and landuse mapping

One of the principle criteria in studies on vegetation mapping is the homogeneity of the stand. This is easily accomplished by selection of study points in those photo patterns, where the density of cover is uniform [16]. Also, vegetation types have been found to be reasonably well correlated with habitat attributes, notably landforms and soils, causing this relationship to be used for broad scale mapping [17, 18].

Among all the landscape features, major landuse mapping is the easiest to accomplish from aerial photos, which results in an immense reduction in the ammount of necessary field work. Agricultural lands are distinguishable by their checkerboard pattern and tonal characteristics. Often it is possible to distinguish between cropped and fallow lands as well [19, 20]. Lands not under cultivation (waste lands, according to revenue classification) constitute the other major category of landuse in the region. Sen [21] has shown successful mapping of these lands even at subtype level, as sandy waste, rocky waste, gravelly waste and saline waste.

Integration of resource survey for development planning

Multidisciplinary surveys provide a great deal of useful data but because of their disjointed nature it is not in a form easily 'digestible' by planning and developmental personnel. To overcome this handicap, the survey data are integrated to develop resource units that are fairly homogeneous in their land attributes, management needs and resource potential. Each of these units is described for its present use and management of resources, land productivity, scope for upgrading production and suggested treatments for their realization. One of the outputs from this exercise is a proposed land management plan, an illustration of which appears in Figure 1. As of now, 87,000 square kilometers or 44% of the area has been surveyed using this technology.

Satellite multispectral data in resource survey

In the past few years a number of feasibility studies have been undertaken to harness the potential which can be generated by the satellite sensing platforms, primarily Landsat. A brief account of these follows.

Visual analysis of Landsat images

Shankaranarayan and Singh [22], while working on the various bands of the scene covering the middle Luni Basin, found that Band 5 provided a more spectacular view of topography and slope while Band 7 proved ideal for de-

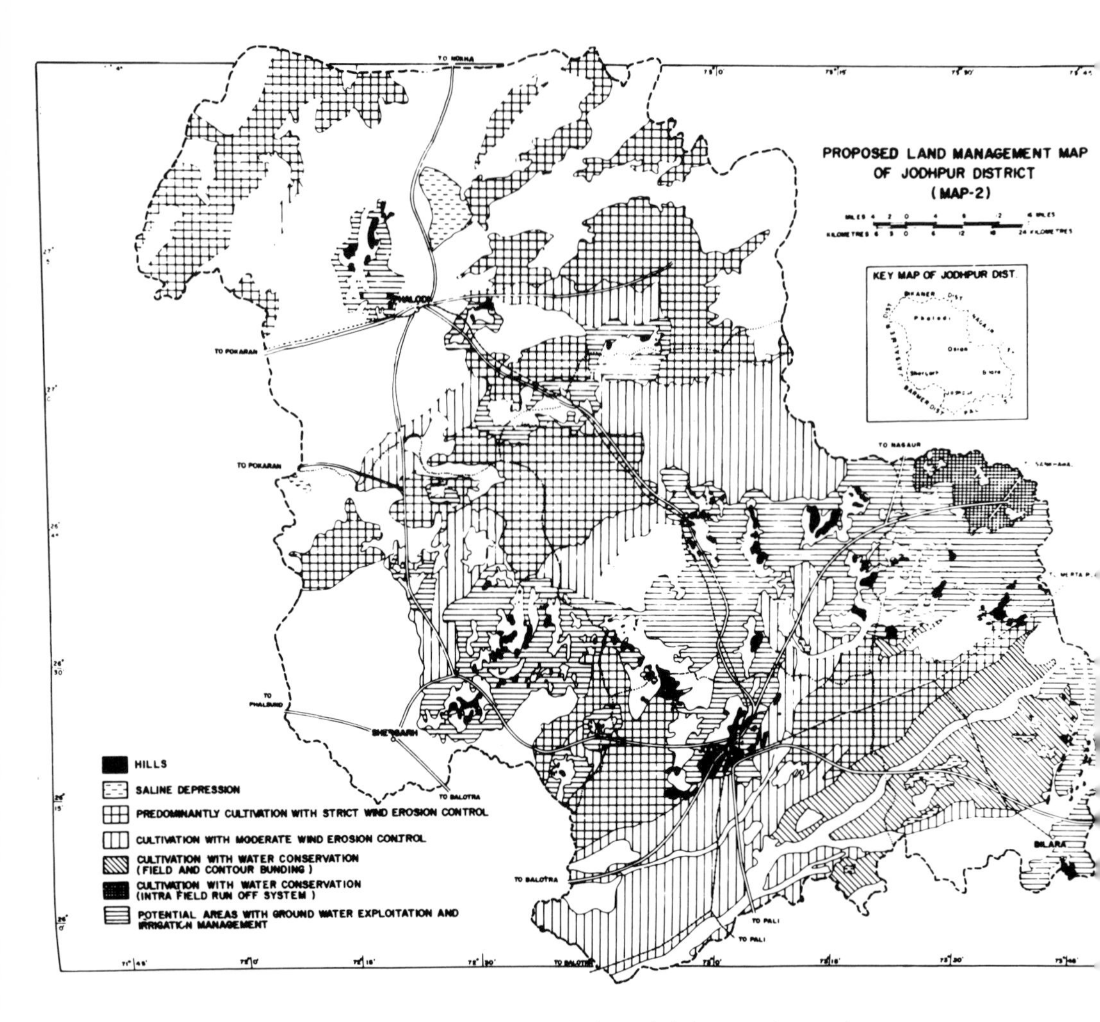

Fig. 1. Aerial photointerpretation has played a key role in integrated natural resources survey, one of the outputs of which is the proposed land management map. The areas left blank are for pastoral or silvi-pastoral land uses.

tecting the joints and fractures in the rocks. A very salient feature was the appearance of granite and rhyolitic rocks as well developed circular features, against the elongated shapes of sedimentary rocks. Such information could not be extracted from earlier studies either by ground survey or aerial photographs. Such unique observable features were taken advantage of, to easily delineate between fourteen landform units and associated soils. Likewise, surface water resources could be easily observed in all the spectral bands of the image. Among the four bands, Band 7 was ideally suited to sharply demarcate the surface water bodies from surrounding areas (Table 1), whereas Bands 4 and 5 provided a qualitative assessment of sediment and depth respectively. Sand dunes are visible with utmost clarity in Band 7. The reflectance properties of the dunes are mainly responsible for the detection of the shape and extent of the dunes. Landsat imagery of the dry, cold season was ideal for distinguishing between stabilized and active dunes.

Dhir et al. [23] showed that with Band 7, the dominant coarse loamy Camborthids could be recognized with ease, and vegetation cover was found to be non-interfering, though surface soil moisture variation appeared to do so. Hard pan soils were identifiable more by the manifestation of associated features than by the soil characteristics themselves. Fine loamy Camborthids could be recognized at a time when the land is devoid of much of its vegetation cover. Most of the saline areas could be recognized, but those in the southeastern tracts were largely inseparable from the adjoining shallow, eroded soils. For these, Band 5 images from during the monsoon months were more satisfactory.

Table 1. Temporal changes in the surface area of tanks, based on the sequential Landsat images of one scene (Area in km^2).

S.M.	Years and scene ID	Nayagaon Tank	Sardar Samand Tank	Kharda Tank	Hemawas Tank
Dry cool season					
1.	Jan. 10, 1973 (1171–05092)	0.24	4.23	0.85	1.95
2.	Feb. 14, 1975 (2023–04574)	0.36	5.29	2.84	0.39
3.	Jan. 16, 1977 (2725–04441)	5.62	45.90	17.39	12.55
Dry hot season					
4.	Apr. 3, 1976 (2437–4532)	3.92	45.41	18.21	15.00
5.	Mar. 11, 1977 (2779–04420)	4.01	38.72	14.17	4.32
6.	May 22, 1977 (2851–04382)	2.71	33.42	10.30	0.42
Wet season					
7.	Sept. 24, 1972 (1063–05090)	1.68	8.63	9.62	12.28
8.	Sept. 9, 1973 (1423–05081)	10.56	80.00	24.40	38.94

Table 2. Range biomass in test sites from ND 6 computer printout.

Gray level Symbol	Subsurface number and number of pixels				Mean	Total area (Ha)	Total biomass (Kg)
	1	2	3	4			
Younger alluvial plain							
	12	5	1	1	4.75	1.9	36.1
	102	112	116	40	92.50	37.0	14,060.0
	148	176	166	140	157.50	63.0	35,910.0
	33	26	34	78	42.75	17.1	12,996.0
	12	0	3	28	10.75	4.3	4,085.0
Blank	13	1	0	13	6.75	2.7	3,078.0
Total						*126.0*	*70,165.1*
						Average yield/ha 556.8 kg	
Pipar soils							
	0	1	1	0	0.50	0.2	23.2
	65	45	49	67	56.50	22.6	5,243.2
	156	157	119	187	154.75	61.9	21,541.2
	75	104	112	63	88.50	35.4	16,425.6
	4	12	36	3	13.75	5.5	3,190.0
Blank	0	2	4	0	1.50	0.6	420.0
Total						*126.2*	*46,843.2*
						Average yield/ha 371.2 kg	
Chirai soils							
	1	0	0	1	0.50	0.2	26.6
	67	34	95	94	72.50	29.0	7,744.0
	191	168	156	158	168.25	67.3	26,852.7
	57	108	59	64	72.00	28.8	15,321.6
Blank	4	10	0	3	4.25	1.7	1,130.5
Total						*127.0*	*51,045.4*
						Average yield/ha 401.9 kg	

Visual analysis of Landsat false color mosaic

Visual interpretation of a Landsat false color mosaic of Rajasthan in conjunction with ground truth data led to the accurate identification, delineation, and mapping of nine geomorphic units. It enabled clear demarcation of two major drainage systems, namely the Ghaggar drainage system and the Luni Jawai system. Water bodies like Sambhar Lake were prominently observable in dark blue color used for accurate delineation. Vegetation was not easily observable, except that the dark red colored patches on the Arvalli hills represented the vegetation in the colored mosaic. The spectral signature was inadequate to permit direct interpretation.

Digital analysis of M.S.S. spectral data

Shankaranarayan and Singh [22] used the analysis of computer compatible tapes in the quantification of the range biomass. Employing a vegetation parameter which utilizes the normalized difference (ND 6) of the sun-angle, corrected rediance values from the red band (Band 5) and one near the infrared band (Band 6), they found it possible to estimate the range biomass in three soil series identified from ground data in the middle of the Luni Basin, namely a) 556.8 kg/ha in alluvial soils; b) 401.9 kg/ha in Chirai soils and c) 371.2 kg/ha in Pipar soils (Table 2). Such results, when examined in conjunction with the repetitive coverage which is offered by satellites, would be of predictive value in the monitoring of temporal changes in biomass of rangelands, which is important information for use in adjusting stocking rates of animals in grazing management. They further showed the amenability of the data for mapping the distribution of sand dunes, interdunes, fluvial and aeolian landforms, alluvial plains and water bodies in quantified terms.

Use of the interactive multispectral image 100 systems was investigated for two sub-scenes, namely Subscene I (Bhetnada-Mori Manana-Kankani sector) and Subscene II (Balesar sector) of the middle Luni Basin, which were examined in the Image 100 system. Results showed that it is possible to classify subscene I into eight spectral themes, by grouping the pixels of similar reflectance characteristics, brightness values and color. These could then be quantified in terms of pixel numbers, areas, and the percentage of the total scene area. Following the same approach, the subscene II (Balesar sector) was classified into six themes which could be quantified in terms of pixel numbers, areas and percentage of total scene area.

The general correlation of spectral classes to the integrated land units would imply that digital mapping of units may precede field work in land surveys and act as a guide to field sampling and detailed descriptions of terrain units as well as measuring the location area and the extent of each unit.

Remote sensing in reconstruction of prior drainage network

An important finding from the viewpoint of reconstructing the evolutionary history of the arid zone, as well as from pragmatic considerations has been the recognition and mapping of prior drainage networks in the region. Ghose [24, 1] from a study of aerial photos observed relics of a large number of prior stream channels. This drainage system was active from a long time and was responsible for major sedimentation in the region but got disorganized when aridity set in due to the paucity of runoff or to choking by aeolian sands.

It has been seen that there is a subterranean flow through these prior streams

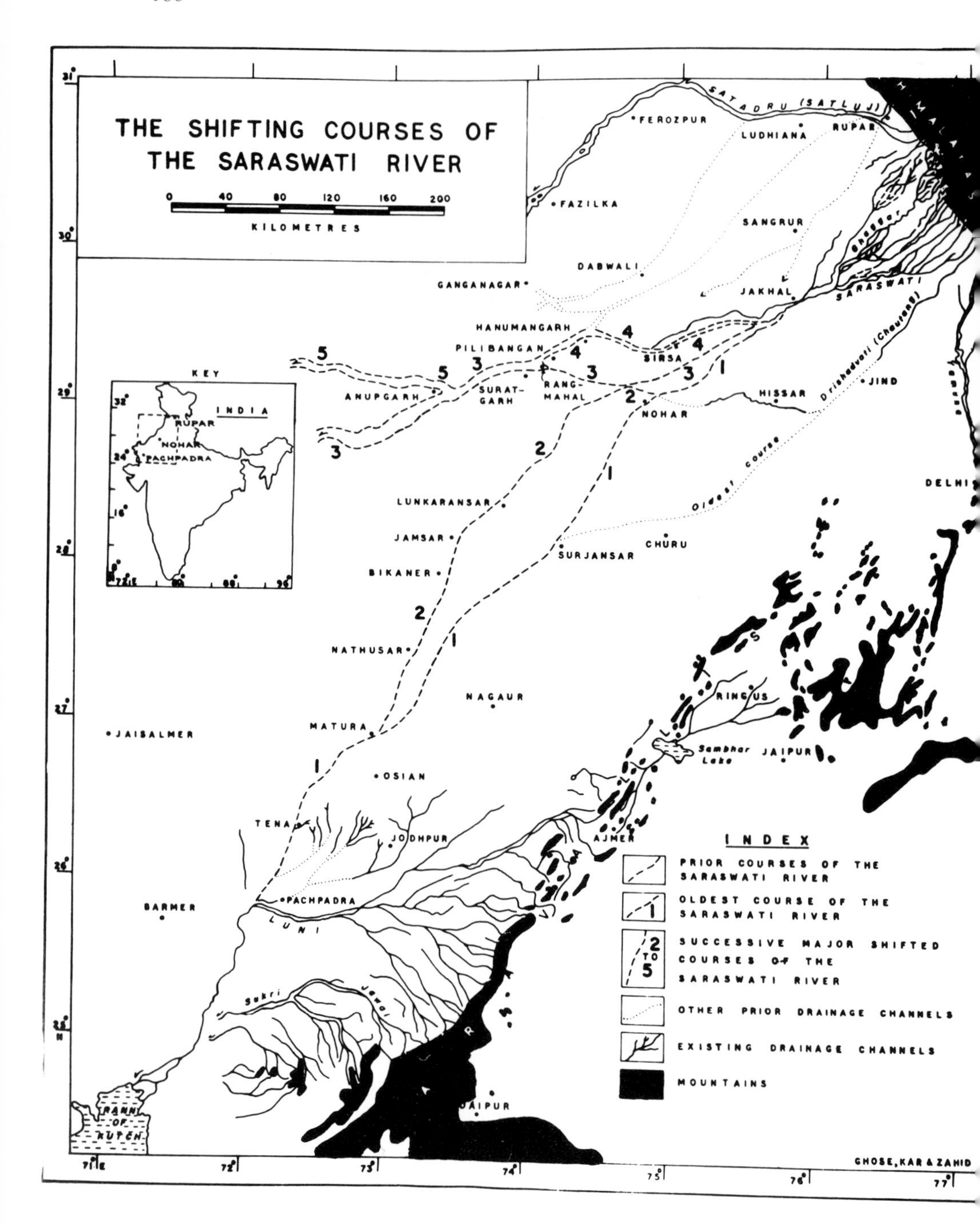

Fig. 2. Prior courses of the lost Saraswati River recognized with the aid of Landsat imagery and aerial photographs [25].

which makes the latter a prospective area for ground exploration. The finding on prior drainage networks has helped in explaining the distribution of major salt basins in the region. Continuation of this work based on aerial photos and Landsat imagery has since helped in establishing (Fig. 2) the prior channels of the mighty Saraswati River – that at one time flowed through this tract [25].

Remote sensing in desertification studies

As stated earlier, the arid zone of India, including Rajasthan, is distinguished by its prolonged and intense use of natural resources. Even the most arid parts bear the brunt of extreme human pressure. In the zone where rainfall is above 250 mm, arable farming is the dominant landuse, in more desiccated tracts it is open grazing. Arable farming on sandy, erodible soils without adequate control measures has lead to a problem of surface instability and accelerated sand movement. Overgrazing in the pasture lands has greatly diminished the perennial vegetation cover. Cutting of trees and shrubs for fuel and over-exploitation at places of groundwater are the other contributing factors. Therefore, under both major landuses, some deterioration of environment is apparent. The presence of barchans in the immediate vicinity of settlements, degraded pasture lands, and secondary salinization in irrigated lands are some of the vivid manifestations. Yet according to present assessment, the rate of desertification processes fortunately is slow. Furthermore, it does not have consistency over time. Whereas during a series of droughts, there is a conspicuous escalation, during wet years, the process gets reversed. Therefore, with the knowledge of the large inter-annual variations, appreciation of this gradual trend of desertification is nothing short of a challenge. Remote sensing tools, because of their unique advantages can play an important role in the monitoring of this process as well as the recognition of the hazard of vulnerability.

Mapping vulnerability to desertification

Remote sensing techniques have been of great value in monitoring desertification. The tonal and textural variations in images has helped to detect, delineate and map desertification hazards such as wind deposition and deflation, water erosion, and natural and man-induced salinity. Thus, 30% of the scene area was affected by depositional hazard; 4% by wind deflation and 10% by water erosion and hazards of wind deposition.

Natural salinity hazards were observed in the vicinity of Thob, Pachbadra, Sanwarla, Harji and Manador villages covering 3% of the scene area. Man-induced salinity was observable in medium to heavy textured alluvial plains in the eastern part of the scene, affecting about 8% of the scene area.

Based on the light and dark tone pixels in the Band 5 computer printout map of a subscene in the middle Luni Basin, it was possible to delineate areas with dark pixels (0–65 brightness value) covering an area of 3.4 km^2 which are considered to be the least vulnerable to desertification. Light to dark gray pixels (66–69 brightness value), involving an area 6.1 km^2, indicate areas moderately vulnerable to desertification. The bright pixels (70–83 brightness value), occupying an area 4.7 km^2, constitute areas highly vulnerable to desertification.

Hydropeilous desertification

The manmade decision to divert the excess water flow in Ghaggar into 18 interdunal depressions of various sizes involving an enormous quantity of water (0.73 million acre feet) has had a disastrous effect due to the seepage of water from stored depressions into fertile agricultural lands (792 hectares in Baropal), which have now been invaded by aquatic weeds like *Typha anqustata* and *Arundo donax*. The false color composite scene of this area showed the water-logged areas in mottled dark blue color, which could be mapped with accuracy from surrounding areas.

Environmental monitoring program

Encouraged by the results obtained by feasibility studies based on Landsat data, sites in the arid regions of four northern States, namely Punjab, Haryana, Rajasthan, and Gujarat have been chosen for repeat observations, due to the presence of critical indicators of desertification in these areas which constitute monitoring proper. Within this framework, the following permanent sites have been selected for further quantitative monitoring.

I. Sand activity and movement
  a) Mahajan village
  b) Mandloi-Satlana
  c) Kadana-Naurangwas

II. Water logging and salinity
  a) Baropal-Manaktheri
  b) Tibi
  c) Nohar-Rawatsar
  d) Lunkaransar

III. Enlargement of rock surface
  a) Saradhana
  b) Sendra
  c) Bar-Beawar

To summarize, aerial photointerpretation has consistently proved to be a powerful tool in natural resources survey. Its potential in environmental monitoring is also obvious, but the cost involved in repetitive coverage is a serious constraint. Therefore, what is envisioned is a repetitive low altitude aerial photography of test sites only with large area extrapolation based on satellite sensed data. Large synoptic view, repetitive coverage, multi-band sensing and amenability of data to statistical analysis are great advantages with the Landsat products. Enhanced spatial resolution and additional spectral bands in future generations of the satellite program will further enhance the utility of satellite-based sensing platforms.

References

1. Ghose B: The genesis of the desert plain in the central Luni Basin of western Rajasthan; J. Indian Soc. Soil Sci. 13:123–126, 1965.
2. Ghose B, Singh S: Analysis of landform of Saila Block from aerial photographs. Ann. Arid Zone, 4(2):207–216, 1965.
3. Pandey S: Geomorphology of Jalor and adjoining region. Annals Arid Zone: 4(1):74–83, 1965.
4. Singh S, Ghose B: The application of aerial photointerpretation in geomorphological surveys. The Deccan Geogr. 4(1):1–13, 1969.
5. Pandey S, Singh S, Ghose B: Orientation, distribution and origin of sand dunes in central Luni Basin. Proc. Symp. Probl. Indian Arid Zone: 84–91, 1961.
6. Vats PC, Singh S, Ghose B, Kaith DS: Types, orientation and distribution of sand dunes in Bikaner district. Geogr. Observer 12:69–75, 1976.
7. Ghose B, Pandey S: Quantitative geomorphology of drainage basins. J. Indian Soc. Soil Sci. 13:123–126, 1963.
8. Ghose B, Pandey D, Singh S, Lal G: Quantitative geomorphology of drainage basins in central Luni Basin. Zeitech fur Geom. 11(2):146–160, 1967.
9. Abichandani CT: Use of aerial photographs in soil survey of arid zone of Western Rajasthan. Annals Arid Zone 4(2):172–184, 1965.
10. Dhir RP: An approach to use of aerial photographs in small scale soil mapping. J. Indian Soc. Photo.-Int. 2(1):13–18, 1974.
11. Dhir RP, Singh N: Nature and incidence of soil salinity in Pali Block. Annals of Arid Zone 18:174–180, 1979.
12. Kolarkar AS, Dhir RP, Singh N: Characteristics and morphogenesis of salt affected soils in southeastern arid Rajasthan. J. Indian Soc. Photo-Int. and Remote Sensing 8(1):31–41, 1980.
13. Chatterji PC: Classification of groundwater potential zone. Proc. Nat. Acad. Sci. India, Series B 34(IV):402–418, 1964.
14. Chatterji PC, Singh S: Geomorphological studies for exploration of groundwater in Rajasthan desert. Proc. Indian nat. Sci. Acad. Series A 46(5):509–518, 1980.
15. Chatterji PC: Application of aerial-photointerpretation in hydrogeomorphological investigation. Proc. Symp. 'Remote Sensing for Hydrological, Agricultural and Mineral Resources', Space Application Centre (ISRO) Ahmedabad: 121–126, 1977.
16. Raheja PC: Aerial Photointerpretation for survey of natural resources. Proc. Probl. Indian Arid Zone: 459–464, 1964.
17. Satyanarayana Y, Dhruvanarayana V: Use of aerial photos and surveying groundwater and

vegetation resources. Toulouse, France: Toulouse Conf. UNESCO, 1964.
18. Gupta RK, Abichandani CT: Air photo analysis of plant communities in relation to adaptic factors in arid zone. Proc. Symp. Recent Adv. in Eco. 18, 1967.
19. Sen AK: Landuse mapping by aerial photointerpretation. In: CAZRI Desert Ecosystem, Jodhpur, India: Central Arid Zone Research Institute, 1977a, pp 85–100.
20. Sen AK: Aerial Photointerpretation to analyze landuse patterns of sand dunes. The Deccan Geogr. (4):346–352, 1977b.
21. Sen AK: Land utilization mapping to map wastelands using aerial photointerpretation. Indian Nat. Sci. Bull. 44:67–71, 1972.
22. Shankaranarayan KA, Singh S: Application of Landsat data for natural resource inventory and monitoring of desertification. Training Report, Visiting Intern. Science Programme, Sioux Falls, South Dakota: Remote Sensing Institute, 1979.
23. Dhir RP, Kolarkar AS, Singh N: Recognition of arid zone soil features from Landsat imagery. J Indian Soc. Photo-Inst. 6(1):9–13, 1978.
24. Ghose B: Geomorphological aspects of the formation of salt basins in western Rajasthan. Proc. Symp. Probl. Indian Arid Zone: 79–83, 1964.
25. Ghose B, Kar A, Qureshi, Z.H.: The lost course of the Saraswati River in the Great Indian Desert – New evidence from Landsat imagery. Geog. J. 145(3):446–451, 1979.

Authors' address:
Central Arid Zone Research Institute
Indian Council of Agricultural Research
Jodhpur 342003, India

10. Landsat surveys of southeastern Arabia

John R. Everett, Orville R. Russell and David A. Nichols

Introduction

Arid lands in general, and the southeastern Arabian Peninsula in particular, are nearly ideal environments for applying remote sensing technology to environmental and natural resource problems, both in terms of operational conditions and the yield of beneficial information. The dry climate makes it relatively easy to obtain cloud-free imagery under excellent atmospheric conditions, particularly during the winter months.

The same dry climate that fosters acquisition of excellent images presents a number of challenges to human survival and has limited the size and distribution of the population of the area. This, in turn, has limited knowledge of the nature and distribution of natural resources in the region.

Taken together, the conditions of climate and the general paucity of knowledge create a context in which remotely sensed data, particularly Landsat data, can make a large contribution. The contribution rests not only on the information content inherent in the data, but in a larger sense on the new and unique perspective that the data provides.

There are at least three valuable elements of this new space perspective. Firstly, the large area covered by a single image allows one to interrelate features seen on the image to surface conditions that may be separated by substantial distances. Specific 'ground truth' available for a limited area on the Landsat scene can be extended to other areas that are imaged similarly, and previously unrecognized relationships among separate features may become obvious.

In addition, the large area covered by each Landsat image is produced under uniform atmospheric and lighting conditions allowing the confident perception of subtle variations in tone, color and texture on the image. This may be important for understanding the nature, variation and distribution of various natural resource complexes. Many of these variations are so slight or gradual that they escape notice on the ground or in conventional aerial photography.

Finally, this radical new perspective compels us to view and consider the Earth in a new and different way. This generates new ideas and insights that may

El-Baz, F. (ed.), Deserts and arid lands. ISBN: 90-247-2850-9.

challenge some existing precepts and assumptions. This, hopefully, engenders a more complete and integrated understanding of an area.

More immediately useful than the specific information derived from the Landsat images are observations on the approach to various resource complexes, the methodology used and the role of available information and field work in providing a realistic context for interpretation. There are several recurrent themes evident in all the work. The ability to view large areas at one time is important in each of the studies discussed, especially in regional tectonic studies and exploration efforts. Detection of subtle tonal differences is important for locating fractures, particularly in sand covered areas, and it is critical in exploration for water or hydrocarbons. Detection of shapes, breaks, or small zones of vegetation contribute to all efforts. This new perspective permitted by the space imagery provides a special advantage unique to this tool.

During the past several years we have had an opportunity to examine much of the southeastern portion of the Arabian peninsula for a variety of purposes. Some of the information derived is proprietary to specific clients. This condition and the simple limits of time and space dictate that this chapter is not an encyclopedic compedium of information on the southeast Arabian Peninsula that may be derived from Landsat imagery.

Each of the following discussions, though general in content, indicates the surprisingly detailed and profound insights that Landsat imagery can provide. It is clear that these insights establish a sound basis for further work that can be efficiently focused on specific areas.

The work reported here represents a few man years of effort and was completed in less than two years. The point is that for each of the subjects discussed (albeit briefly) conventional approaches would require, and have required, tens of man years of effort and produced less comprehensive or reliable results. These considerations are particularly important to countries that have only relatively recently taken the responsibility for managing their resources.

Subsequent discussions cover various parts of the United Arab Emirates, the Sultanate of Oman and eastern Saudi Arabia (see Fig. 1). All of the studies relied on Landsat imagery as an initial data source and included field work and literature research as critical elements in understanding and refining the observations made from the space imagery. We gratefully acknowledge the hospitality of the people of the countries involved.

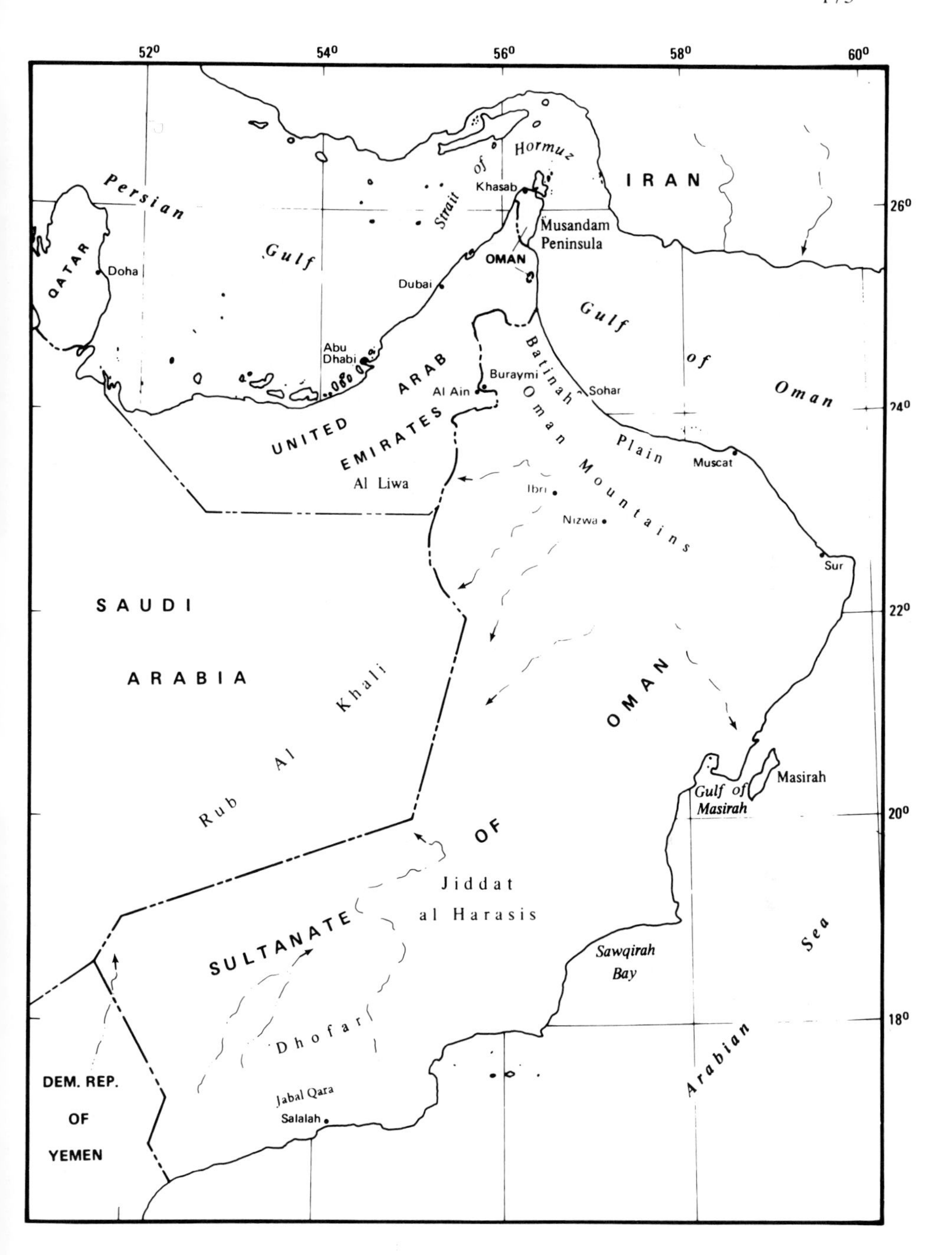

Fig. 1. Map of the southeastern Arabian peninsula. All portions of this area are discussed in the various sections of this paper.

Land use and terrain analysis in northern Oman

Introduction

Landsat imagery is a valuable tool for land evaluation because it provides an inexpensive view of the land surface at a regional scale. Users and analysts must be aware, however, of the limitations of the Landsat land evaluation approach. A normal tendancy is to try to attain unrealistic levels of detail in the interpretation that are beyond the capabilities of the system. It should be remembered that interpretation is an analysis of existing knowledge applied to an area on the basis of its appearance on the Landsat image. Therefore, one must acknowledge that erroneous interpretations will occur. These errors can be kept to an acceptable minimum if the limitations of the Landsat system are not exceeded and an effective program of data gathering is applied in conjunction with the interpretation effort.

A regional classification of a landscape can be directed towards any specific discipline desired, such as agronomy, geology, agriculture, etc. Regardless of which end use is intended, the ability to view a large land area as a synoptic entity allows the analyst to study cause and effect relationships and to draw area-wide conclusions based on all ecological systems at work.

In arid lands the importance of the ecosystem is crucial because the line between potentially productive land and 'wasteland' is very narrow. One need only look at the causes of the American 'Dust Bowl' in the 1930's and the encroachment of the desert in the African Sahel to see the effects of development activities that failed to consider the fragile ecosystems they effected. This is not to minimize the importance of micro-scale analysis. Any land development project requires careful consideration of local landscape parameters as well. Too often, however, local development projects (and not only those in arid lands) fail because the regional parameters were not considered in the planning process.

The purpose of this study was two-fold. The first part of the effort was to show terrain features and land use activity as it presently exists. This would provide a regional framework to determine locational patterns and relationships which are important to resource managers and land planners. The second aspect of the effort was to provide a basis for a future multi-disciplined data bank for land planning. This land evaluation study was undertaken to demonstrate the type of information which could be effectively derived using Landsat imagery as a data base. The study area was located in northern Oman and included coastal plains, inland plains, mountains, and an erg desert as principal landform types. This project represents a preliminary investigation so the opportunity for field surveys and literature review was minimal.

Methodology

Landsat imagery was used to classify landscapes on a regional basis, using the concept of repetitious environments. The use of Landsat imagery enabled the land analyst to group common characteristics of soils, landforms, land use, drainage, and vegetation into areas containing similar characteristics. This would be impossible or ineffective by means other than using the satellite image synoptic perspective.

The study consisted of the analysis of Landsat imagery covering the Batinah Plain, the Oman Mountains, and a portion of the Wahiba Sands. Figure 2 shows a portion of the area studied. The black-and-white print of spectral Band 5 (0.6–0.7 μm wavelength) does not portray the area in the clarity of a color composite image. However, the accumulation plains on both sides of the mountain are prominent as are the few major wadis through which the run-off is channelized. The dark spots distributed along some of the wadis and the dark zone along the shoreline represent cultivated areas.

A brief literature review provided the analyst with knowledge of local conditions, terrain features, and vegetation patterns in the study area. The Landsat scenes were than analyzed for landform, natural vegetation, and land use activities. After this initial interpretation, a field survey team travelled to Oman for a ten day 'ground truth' effort to determine the accuracy of the Landsat analysis and to gather additional data. Upon returning from the field, final terrain feature delineations were made on the Landsat images at a scale of 1:250,000, and a final terrain analysis of the study area was completed.

Landform, vegetation, land use analysis

The landform was found to be the most readily interpretable arid land terrain component viewed from Landsat. Since the vegetation cover is negligible and soil development is poor, the detail of erosional and depositional forces was clearly visible on the Landsat images. Variations in surface material were likewise easily discernible. The progression of land shaping forces was studied and long term trends were also analyzed.

Land use practices were also easily interpretable when the activity was within the resolution limits of the Landsat system. Irrigated agriculture surrounding population centers and major road networks connecting villages and towns are clearly seen on Landsat images.

Natural vegetation patterns in the arid environment of northern Oman were more difficult to interpret. Normal reflectance values of the desert vegetation recorded on Landsat imagery are very low and easily overlooked. Arid land vegetation is usually associated with particular micro-environments rather than

Fig. 2a. Landsat image acquired from an altitude of about 900 km, depicts an area of about $185 \times 185\,km^2$, (or about $3400\,km^2$). This image shows some of the areas of highest potential for groundwater and agricultural development adjacent to Jabal Akhdar.

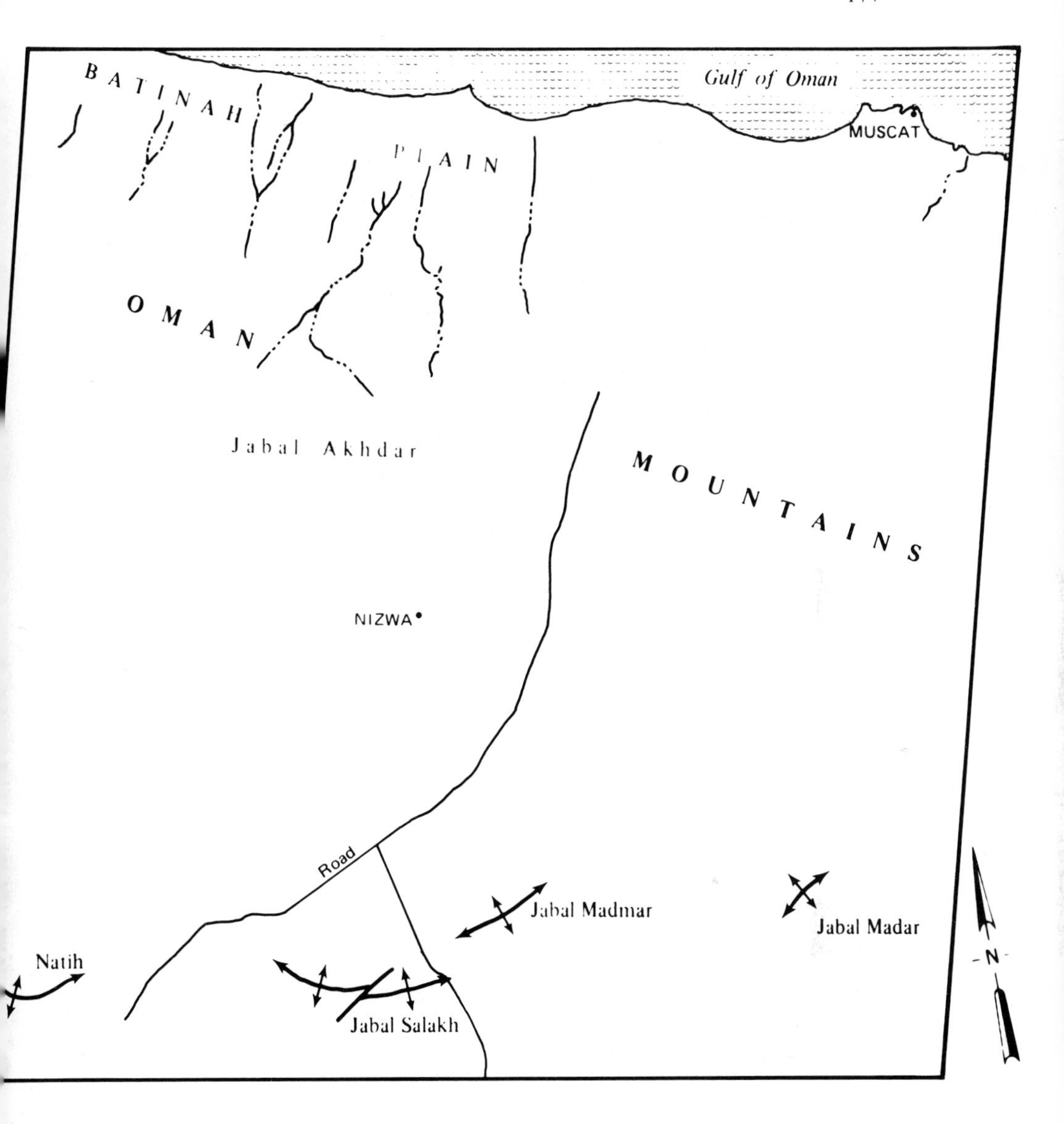

Fig. 2b. Drawing of the main features in the same Landsat image as Figure 2a. In this region the greatest quantities of rainfall in southeastern Arabia occur in this area located on the Jabal Adhdar massif which rises to over 3000 m. This mountain is a large anticlinal feature whose center portion has eroded away, forming a large interior valley surrounded by near vertical walls. Rain falling within this valley flows northward across the Batinah Plain along only two major wadis. The Jabal Akhdar, which is composed of massive limestone, is almost entirely surrounded by ultramafic rocks of the Semali Nappe, an ophiolitic suite that was thrust inland many kilometers before the Jabal was formed. The Jabals mapped at the lower edge of the image are probably diapiric structures, cored with salt and actively 'growing' today.

an area-wide coverage characteristic of more temperate regions. However, the literature review and the field survey enabled the analyst to determine associated vegetation species in these micro-environments and to apply that knowledge to the regional interpretation. In effect, 'knowing where to look' enabled the analyst to interpret vegetation patterns with a higher level of accuracy than otherwise would have been accomplished.

The following section describes the results of the northern Oman terrain analysis. All landforms and land use features described are visible on Landsat imagery. Many details, however, cannot be directly determined from Landsat image interpretation and are the result of literature review and on-site field work. Most of the land shaping processes described do produce recognizable patterns on the image which, once understood, can be analyzed from the imagery. This demonstrates the necessity of an integrated approach to terrain analysis using Landsat imagery, prior knowledge, and field surveys.

Coastal plain

The coastal plain is determined by the influences of the sea on the nature of the landscape. The topography is flat, characterized by marine deposited sand and fragmented shells. Characteristic land forms are active and relic sand bars, lagoons, and sabkhas.

A wave built sand bar exists along much of the Batinah coast. Once emerged, wind action has worked the sand into low coastal dunes. This bar is situated 5 to 10 m above sea level and protects low lying coastal flats behind it. In some cases, the sea has forced gaps in the sand bar (khawrs) allowing inundation of the area behind the bar, forming coastal lagoons. Flood tide deposition in the lagoons eventually creates coastal sabkhas. Coastal sabkhas are relic lagoons which have been silted up to their present height and become inundated only when normal tidal flows are exceeded. Dry sabkhas are usually salt encrusted and provide excellent wind erosion surfaces. Sand from the coastal bar is blown and washed inland, deposited on the sabkhas, eroded (deflated) and redeposited inland [1]. Relic sand bars and sabkhas occur inland of the present 'active' ones indicating an emergent coastline.

Northern Oman's most intensive land use occurs in this area along the sand bars. Agriculture in the coastal plain consists mainly of date palms and citrus trees (limes), both of which are somewhat tolerant of saline soils. Generally, this region does not extend more than five or six kilometers inland from the coast. Increasing population pressure on the coastal plain is placing a greater demand on the available groundwater supply for irrigation and human consumption. As the available groundwater is drawn down, salt water from the Gulf of Oman is being drawn into the groundwater table along the coast. As a result, the agriculture of the coastal plain is characterized by declining crop yields and less favorable soils are coming under cultivation along the coast highway [1].

Accumulation plains

The accumulation plains extend from the coastal plain inland to the Oman Mountains, and from the mountains to the desert dune fields of the Rub al Khali. They are defined by alluvial and aeolian deposition rather than the marine and aeolian deposition of the coastal plain. The accumulation plains of the eastern or seaward side of the mountains differ from those of the interior in material and landform, but geomorphologically they are broadly similar. Close to the mountains the plains are comprised of alluvial sediments deposited by wadis and occasionally sheet flooding. Common alluvial deposition patterns occur with coarse, heavy material deposited in and around the wadi beds, and progressively finer material being deposited further away from the source. Further distance away from the mountains, the alluvial material is interbedded with fine aeolian deposition from the coastal plain or Rub al Khali [1].

Local dune formation occurs where the fine sands are deflated and redeposited, often using tussocks of grass or small shrubs as a base, forming *nebkhas*. In some areas, the alluvial layers are totally deflated, leaving a surface cover of stones and gravel [1].

The Batinah accumulation plain east of the Oman Mountains (see Fig. 1) is characterized by wadi terraces and single channel wadi beds. These terraces are formed by cementation of old wadi fans, which are now deeply dissected by the wadi channels. The surfaces of these old terraces are usually deflated, leaving a landscape of boulders and gravel [1]. This process strongly suggests that the Batinah coastline is emergent.

The inland accumulation plain is characterized by younger, generally uncemented, alluvial fans. Wadi beds are wider and often braided in appearance. Many of the fans are extensive (as much as 20 km across) and gradually blend into the surrounding landscape or one another with poorly defined lower boundaries.

Agriculture is localized in the wadi beds and irrigated by the *falaj* system. Primary production is in date palms, limes, mangos, almonds, alfalfa, sorghum, and onions. Secondary crops include melons, radishes, garlic, tomatoes, peppers, papaya, and pomegranites [1].

Mountains

The Oman Mountains are steeply sloping, barren of vegetation and have little soil development. The range is between 50 and 70 km wide and trends NW-SE, parallel to the Gulf of Oman coast. The mountains provide the bulk of source material for the accumulation plain, and orographic effects of the Jabal Akhdar provide a relatively significant recharge process for the available groundwater in northern Oman.

Agricultural land use occurs in spring fed mountain wadi beds and 'hanging garden' terraces. Principal crops include date palms, limes, and some sorghum and vegetables. Local livestock grazing occurs on some of the higher, gentler slopes where the grass cover permits [1].

Desert dune fields

The desert dune fields are broad areas of sand dunes and sand sheets bordering the inland accumulation plain. In some cases, the boundary is quite distinct and somewhat arbitrary in others, due to the mixing of finer, easily transported sediments of the accumulation plain with the blowing sands of the dune fields.

Vegetation

As stated previously, natural vegetation in arid lands is usually extremely difficult to interpret directly from Landsat imagery. Our study technique of using existing data and field surveys enabled the analyst to interpret landscapes where vegetation was likely to occur in order to improve the accuracy of the vegetation analysis. By being familiar with the ecological requirements of the vegetation, and confirming this knowledge in the field, the analyst was able to 'see' vegetation patterns on the imagery which normally would not be interpretable.

Natural vegetation on the coastal plain is limited by high salinity soils and human impact. Most, if not all, of the arable land in the coastal plain is already under cultivation, thereby eliminating much of the natural vegetation. What vegetation remains in the coastal plain area is limited to the old sand bars and dune areas where there is little soil development. The resulting vegetation pattern is low scattered shrubs of salt tolerant species, typically *Heliotropium* spp., and *Haloxylon salifonicum*. The grass cover is light and scattered, dominated by *Zygophyllum* spp., and *Calotropis procera* [1].

The accumulation plains can be broadly classified as having a desert parkland type vegetation. The parkland environment is characterized by scattered trees, three to six meters high, with little or no grass cover. Acacia trees are the dominant vegetative feature most commonly represented by *A. tortilis* and *A. ehrenbergiana* [2]. The phraetophytic acacias follow run-off channels and almost imperceptible depressions to make use of the available groundwater. Locally, dense stands of *Acacia nilotica*, *Ziziphus spina-christi*, and *Prosopis spicigera* occur where adequate groundwater is available. Localized occurrences of *Salvadora persica* are common on low dunes and nebkhas, and *Haloxylon saliconicum* frequently occurs on sandy, more saline soils [2].

Vegetation in the mountains (exclusive of wadi beds) is extremely difficult to classify due to its irregular occurrence. Much of the mountain surface is barren rock devoid of soil and vegetation. At lower elevations (less than 1,300 m) where the soil is adequately developed for vegetation, a shrub, *Euphorbia larica*, is a common occurence. Below 1,000 m *E. larica* is often associated with *Acacia tortilis* and *Gaillonia* spp. is associated with it at above 1,000 m. Other species, such as *Ziziphus spina-christi* and *Acridocarpus orientalis*, occur in this region but are primarily confined to ravines and other protected habitats [2].

Existing literature suggests a vegetation zone association of *Reptonia mascatensis* and *Olea africana* between 1,350–2,300 m on Jabal Akhdar. This association reportedly reaches its fullest development in steep sided ravines on the north face of the mountain, sometimes assuming the character of a localized forest (closed canopy) [2]. Our field effort routing did not allow confirmation on this zone, but there is no recorded evidence to cast doubt on its existence.

A final vegetation zone occurs at approximately the 2,000 m elevation contour. At this elevation (respectful of soils and water requirements) there is a notable increase in grass cover, and well developed *Juniperus macropoda* dominates exposed surfaces. The grass association (perennial tussock formations, with clumps approximately $\frac{1}{4}$–$\frac{1}{2}$ m apart) consists primarily of *Heteropogon contortus* and *Cymbopogon schoenanthus*. *R. mascatensis* and *O. africana* exist in association with the junipers, but in a less developed state unless in a ravine or other protected area [2, 1].

Natural vegetation within the mountain wadi beds is somewhat azonal in character due to the habitat protection afforded by the ravines, and often more favorable soil and water conditions. Portions of the wadi beds which are covered with rocks and boulders are usually devoid of vegetation. Loose gravel cover with a bottom alluvial soil layer yields abundant *Nerium oleander* trees in relatively high and dense stands. Increasing distance up the sloped sides of the wadi bed is characterized by *Ficus salicifolia*, *Ziziphus spina-christi* and *Prosopis spicigera* trees – all dependant on the amount of wadi deposited alluvium. *Acacia tortilis* and *Acacia ehrenbergiana* occur on alluvium which has been deposited above the banks of the wadi bed. Grasses associated with local pooled water include *Fimbristylis cymbosa* and *Scirpus litoralis* [2].

Conclusion

Landsat can be a significant aid in arid land studies because of its synoptic coverage. The harsh climate and difficult access of arid lands often makes intensive field surveys quite costly and time consuming. By using Landsat imagery as a guide, field survey efforts can be directed to where they will be most effective. By careful planning and using existing knowledge and expertise, large areas of arid land can be effectively studied for a data base which can then be expanded as planning needs require.

Groundwater studies

Introduction

Topographically, the southeastern Arabian Peninsula is characterized by mountains to the northeast along the Gulf of Oman and to the southeast in the Dhufar area (see Fig. 1). A broad valley or interior depression or structural basin 450 km or more across lies between these mountains. This depression heads on the western slope of a low rise west of the Gulf of Masarah, and opens to the west and northwest, perhaps sloping structurally all the way to the Persian Gulf.

Locally in Oman, there are two or three large basins in the depression that receive runoff from the occasional heavier rains that occur in the mountains. Salt flats appear to be present in at least one of the basins and occasionally, as noted on Landsat imagery acquired in 1972, playa lakes are present. Except near the eastern extremity, much of this interior depression is covered with large, complex crescentic sand dunes with intervening flat bare areas that in places are moist and saline.

The Oman Mountains (also known as the Hajar Range) parallel the coast of the Gulf of Oman and extend from the Musandam Peninsula for 600 km, to the vicinity of the village of Sur. Elevations, locally, exceed 3,000 m near Nizwa, but are generally considerably lower. South of Sur, topography is subdued, gently rising from the sea to a low broad divide which separates the interior depression or basin from the Arabian Sea. This positive area extends southward to join the mountains in the Dhofar area north of Salalah.

The United Arab Emirates, excepting that portion that extends across the Oman Mountains, exhibits little topographic relief and is largely covered with sheets of sand or, as in the Al Liwa area, large crescentic dunes.

The climate is hot and arid, with annual rainfall normally not exceeding 50 mm for much of the area, except in the higher mountains. In the higher mountains near Salalah and southwest to Muscat the annual precipitation may be several times this figure.

The area is sparsely populated, except near the areas of higher rainfall and industrialized coastal cities. Population pressures and increased industrialization have created a need for better water supplies to support development and to increase agricultural productivity. Accurate inventories of the natural resources of the area, particularly the hydrological characteristics, are largely inadequate or non-existent.

It is in such an environment that the analysis of Landsat imagery is particularly useful for making a regional assessment of the existing groundwater conditions and for estimating the potential for further development. Multiple coverage of good quality, computer processed, color composite Landsat imagery has proven highly useful for groundwater assessment. The detection of the presence and degree of concentration of phreatophyte vegetation is an important aspect of the analysis, and only with such imagery can the analyst be assured of obtaining maximum resolution and, consequently, the maximum information possible to be derived from Landsat imagery. Although the stereoscopic viewing of Landsat imagery, which is limited to sidelap of adjacent flightlines, is valuable, monoscopic view of a single Landsat frame provides considerable insight into the geomorphology of the area.

From the Landsat image, the regional regime of the drainage system can be observed instantly. Concentrations of vegetation in the higher elevations of the Jabal Akhdar in the north and Jabal Qara in the south are identifiable by the reddish color of healthy vegetation on the false-color imagery. Intense red color identifies irrigated date palm orchards, and other cultivated crops of areal extent of one hectare or larger. Each of these cultivated areas indicates a local water source of reasonable permanence, whether 'hanging-farms', as in the higher elevations of the Jabal Akhdar, or the falaj irrigated communities on the alluvial slopes and plains away from the mountains. By working from the headwaters of a drainage basin and evaluating geological relationships, geomorphic features and the presence of vegetation, both cultivated and natural, one can make an assessment of the shallow groundwater regime.

Dhofar area, south Oman. In the Dhofar area, monsoon storms sweep onshore and precipitation occurs in a rather definite pattern as evidenced by the distribution and sharp vegetation boundaries discernible on Landsat imagery. At Salalah, on the coast, precipitation is reported to be about 200 mm per year (20 cm = 8 in.); however, the density of vegetation on Jabal Qara, the low mountains (750 $\pm$ m) that rise abruptly behind the city, suggests precipitation is substantially greater there.

These mountains are composed largely of calcareous strata with an extensive and intricate drainage system of wadis developed on both the north and south slopes of the mountain. Numerous lineaments are identifiable in the area from Landsat imagery, and they could be indicative of fracturing and/or faulting.

Porosity and permeability caused by fracturing of rocks has been documented by investigators in other areas as contributing to increased well yield. Thus, good groundwater sources could be present in these calcareous strata especially where transected by lineaments. A detailed examination on Landsat imagery of the wadi channels draining to the interior basin from the north flank of the Jabal Qara shows no phreatophyte vegetative growth. Several possible explanations exist for this situation. The alluvial fill of the wadi channels may exceed the maximum root penetration capability of approximately 50 m for phreatophytes to reach water, or the intricate drainage system observed on the Landsat image is largely 'fossil', and represents a previous period of more abundant precipitation, or fracturing in the bedrock may rapidly drain all but the most abundant runoff to the sub-surface. Today, it appears that little runoff occurs on the north slope of the mountains, because most of the precipitation from storms approaching from the ocean falls on the southern or seaward face of the mountains. It is probably the distribution of rain and the porous, fractured nature of the bedrock that contribute to the lack of vegetation growth on the north slope and the probable lack of shallow aquifers. These same factors should contribute to numerous springs on the south flank of the mountains and reliable sources of water.

Jiddat Al Harasis to Gulf of Masirah, Oman. Northeastward from the Dhofar in the Jiddat al Harasis area (inland from Sawqirah and the Gulf of Masirah) elevations are on the order of 150 m. The bedrock in this area is partly limestone of Oligocene age [3] on which virtually no surface drainage exists. Any rainfall rapidly infiltrates this porous rock. There is some suggestion of karst features in portions of the area and sparse vegetation is present in the depressions. Shallow groundwater sources in this area are probably limited, and the water quite hard.

Structurally, a large north-south trending anticlinal feature lies just onshore from the Gulf of Masirah. The upturned, truncated edge of the Oligocene limestone encircles this positive feature on three sides, creating a possible recharge zone for a deep confined aquifer down-dip to the west, if appropriate lithologic conditions exist. Such an aquifer could constitute a mineable water resource.

Northern Oman. The flanks of the Oman Mountains constitute the area of greatest precipitation in the southeastern Arabian Peninsula and the largest area of potential for groundwater development. High elevations and rugged topography characterize the Musandam Peninsula area. The bedrock in this area consists of highly fractured and dissected thick bedded limestone. Numerous small agricultural plots in the mountains can be seen on the imagery, supported no doubt by outflow from fracture systems. The larger areas of cultivation occur in the major wadis where a bedrock barrier has created alluvial filled valleys with restricted downslope movement of groundwater, or where bedrock merely forces the groundwater movement close to the surface where it can be utilized for irrigation.

Much of the runoff from the mountains in the Musandam Peninsula is channelized through three major wadis. One of these wadis flows northward and enters the Strait of Hormuz at Khasab where there is considerable agricultural development. The other two wadis flow south and west and have formed large alluvial fans where they flow onto the flat coastal plain of Ra's al Khaymah, one of the Northern Arab Emirates. Extensive use of the groundwater flow for agriculture occurs near the toes of these fans where the depth to bedrock is more shallow.

South of the Emirate of Ra's al Khaymah, the Oman Mountains are composed largely of ultramafic rocks [4] that have weathered to rugged but less imposing elevations. These rocks are also highly fractured, producing possible avenues of rainfall infiltration for slower and perennial outflow at lower elevations. Communities exist in these mountains, indicating that water sources are present there, but the villages are widely scattered and small.

Considerable runoff from the Oman Mountains is lost in the sand dune area on the western coastal plains of the United Arab Emirates as a substantial growth of natural vegetation can be seen on the imagery in and along sand dunes as far south as Al Ain-Buraymi, the twin cities on the UAE-Oman border.

Elevations in the mountains, southeast of Buraymi, gradually increase to as much as 3,000 m in the highest portions of the Jabal Akhdar (see Fig. 2). Rainfall increases somewhat proportionately to elevation and relatively large quantities of precipitation occur in these areas. Runoff from the mountains is channelized through a few large wadis which offer several possibilities for major development of the surface and groundwater resources. Numerous large communities exist along these stream courses, utilizing the water for agriculture. Nonetheless, substantial quantities of groundwater are being lost to the desert south and west of the mountains and to the Gulf of Oman, on the north.

Several large wadis drain the intermittent runoff from the Oman Mountains to the northeast across a narrow coastal plain. This, the Batinah Plain, parallels the mountains and extends from the UAE-Oman border to near Muscat, the capital of the Sultanate of Oman. A narrow fringe of date palms occurs along much of the shoreline where the saltwater wedge forces the fresh water to the surface. The color of these date palms, as rendered on the color Landsat images, suggests some stress, probably from saltwater encroachment.

Central U.A.E. and Oman. Large, complex, crescentic sand dunes [5] characterize the interior depression which appears to structurally open to the Persian Gulf through Abu Dhabi (Fig. 3). These dunes, which are 100–150 m high, cover only about 50% of the ground surface. The intervening areas are flat bottomed and in

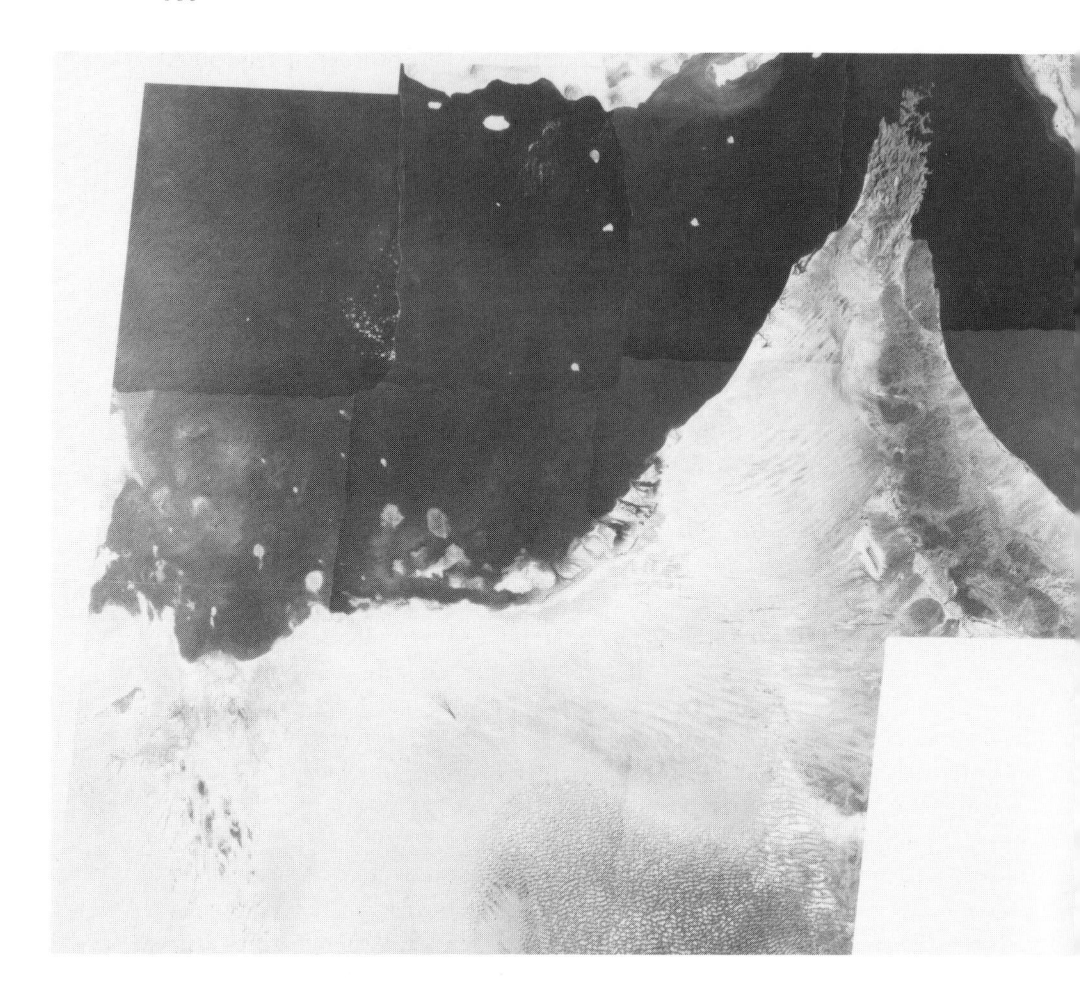

Fig. 3a. This mosaic of eleven Landsat images portrays all of the United Arab Emirates and the northern tip of Oman.

many places show moist ground or sabkha development indicating that the groundwater table is quite shallow. Examination of Landsat imagery reveals no vegetation in the interior of these intradune areas.

An exception to this statement occurs in Abu Dhabi at the interface between the large crescentic dunes and an extensive sandsheet which appears to be encroaching from the northwest onto the large dunes, filling the intradunal depressions. Along this narrow zone, known as Al Liwa or Liwa Hollows, many

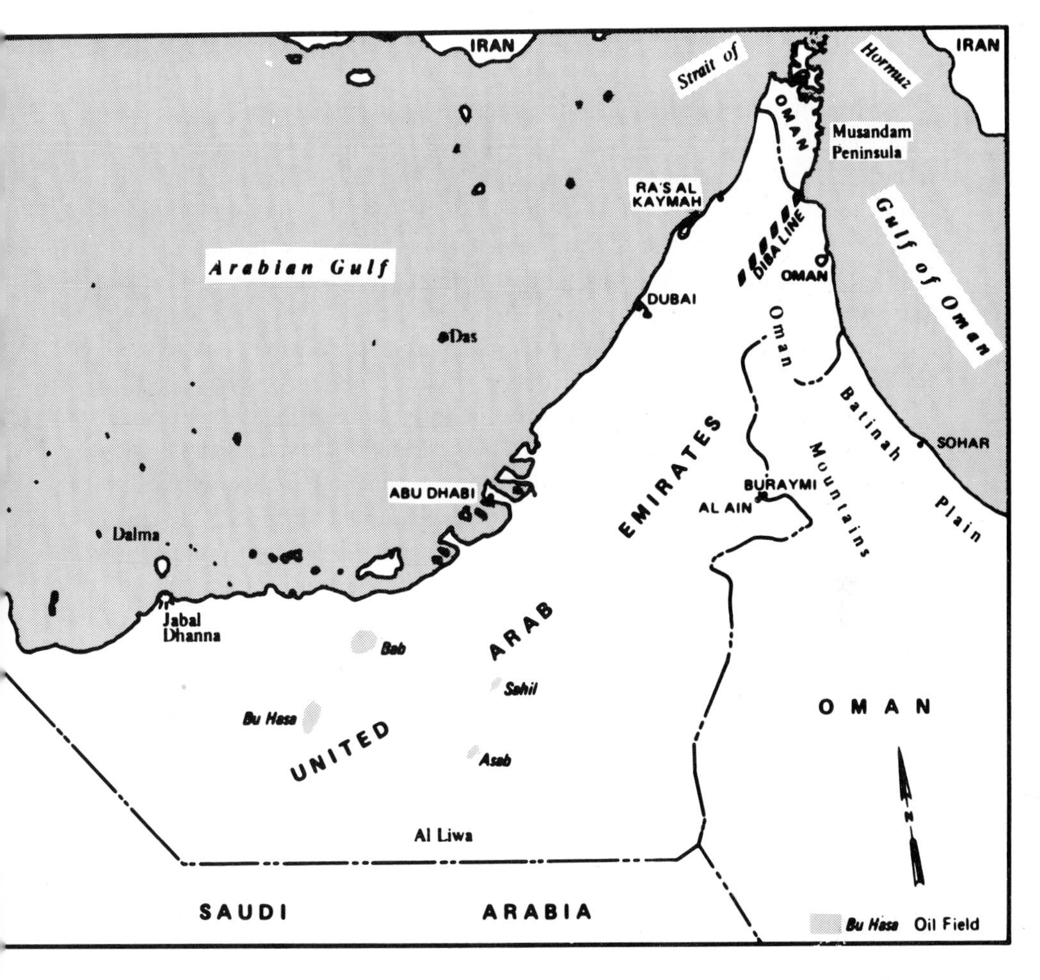

Fig. 3b. Drawing of the same area in Figure 3a. The majority of the islands in the Arabian Gulf are salt diapir structures. Large crescentic sand dunes cover an extensive area in southern United Arab Emirates and Saudi Arabia. Sand is absent from the intradune areas. Although the groundwater is near the surface in these interdune areas, little vegetation exists due to the high salt content of the water and soil.

small agricultural communities thrive at the foot of the dunes in the first unfilled intradunal depression. This anomalous situation is the result of infiltration of the limited rainfall into the topographically higher dune area. The meteoric water appears to percolate downward and laterally, perhaps moving on top of the saline groundwater table, to emergence at the base of the dunes where it is utilized.

Conclusions

The potential for development of the groundwater resources in the southeastern Arabian Peninsula lies adjacent to the Oman Mountains and Jabal Qara. Landsat imagery shows that since 1972 numerous increases in irrigation activities have occurred on the Batinah Plain and near the mountains of the U.A.E. There is potential for much further development in these areas.

One area of substantial agricultural potential that does not appear to have received attention is the south, or interior side of the Oman Mountains. The Landsat imagery indicates that scattered areas of phreatophyte vegetation extend for over 100 km southward from the Oman Mountains over a broad area of braided wadi channels that flow into the interior basin of the Rub al Khali. In this area a detailed examination of the individual wadi channels reveals numerous areas where shallow groundwater occurs and areas where the potential exists for easy capture and utilization. This should be done as far upstream as practical, for much of the groundwater flow that reaches the lower portions of the broad interior depression is probably too saline for utilization.

The foregoing is an initial evaluation of the hydrogeological regime of the southeastern Arabian Peninsula. More specific information about the regional groundwater regime of the area can be derived from a detailed analysis of Landsat imagery combined with aerial photointerpretation and field work. This would provide insight as to where the greatest potential for groundwater development exists, and it would a really focus the application of conventional exploration efforts to optimize groundwater development.

Petroleum exploration

Introduction

The importance of the southeastern Arabian Peninsula to the energy supply of the world cannot be minimized. Although the oil production of the United Arab Emirates and the Sultanate of Oman is much less than some of the neighboring countries, their contribution is substantial. In 1980 their combined annual production was in excess of 730 mln barrels [6].

The area of these two countries and the portion of Saudi Arabia east of 51° E that shows some promise of petroleum production exceeds 520,000 km^2. Current production is derived from about 20% of this area. Much of the remaining area has not been adequately explored, so the potential for increased production is good.

Large portions of the United Arab Emirates and west-central Oman are mantled by alluvial debris derived from the mountains and/or from aeolian sands. Some of the aeolian sands have been shaped into large, complex, crescentic dunes. In much of this area seismic exploration operations for petroleum are difficult and expensive.

United Arab Emirates. Not all of the area of potential hydrocarbon production is on land. Most of the oil from the United Arab Emirates is derived from domal structures in the Persian Gulf, formed as a result of diapiric flowage of the underlying Hormuz salt of Infracambrian Age. The southern end of the Persian Gulf is shallow and a number of these diapiric structures are readily identified on the Landsat imagery as islands or shallow reefs (see Fig. 3). Some of the diapirs are 'breached' and salt is at or near the surface. Normally, these breached structures are no longer sealed and thus contain no hydrocarbons. Structures in which the salt is at considerable depth show little or no surface expression, but are prolific producers of oil.

With Landsat imagery acquired when the water surface is calm, bottom detail can be seen to depths of 30 or more meters. Current channels are readily mappable and several features suggestive of incipient salt domes are visible.

On land, much of the area is mantled by alluvium and sand. In these areas little geological structural detail can be derived from Landsat imagery. However, where the sand is thin or provides only a partial cover, subtle clues to the underlying structure can, in places, be mapped from the imagery. In the U.A.E. this occurs near the Oman Mountains and in the western part of the country as well as in adjacent Saudi Arabia. Near the western Saudi Arabia-UAE border, major east-west trending faults intersect or abut against north-south trending structures. Near the Oman Mountains in Ra's al Khaymah, Umm al Quiwain and Sharjah, anomalous drainage patterns noted on the imagery appear to coincide with faulting. Geologic structure in this area has gained significance by a recent major discovery of oil, the first on-shore discovery in this portion of the Emirates.

A prominent feature in the northern Oman Mountains is the Diba Line. This northeast-southwest trending structural zone marks the present northern limit of the ophiolite rock sequence and is obviously associated with, at a minimum, a major flexure in the Mesozoic strata, and may well also represent a zone of substantial strike-slip movement. A view of regional perspective derived from the mosaicking of Landsat images makes the extension of the Diba Line (Fig. 3) to the producing fields of Sahil and Asab, in Abu Dhabi, look attractive. However, there is no visible evidence on Landsat to support this conjecture.

Oman. There is evidence, from Landsat imagery, of folding and faulting parallel to the Oman Mountains both in the UAE and Oman that are, in part, the result of Miocene diastrophism that affected much of the Arabian Peninsula. Not all of the folding and faulting ceased at that time. Evidence of Recent fault movement

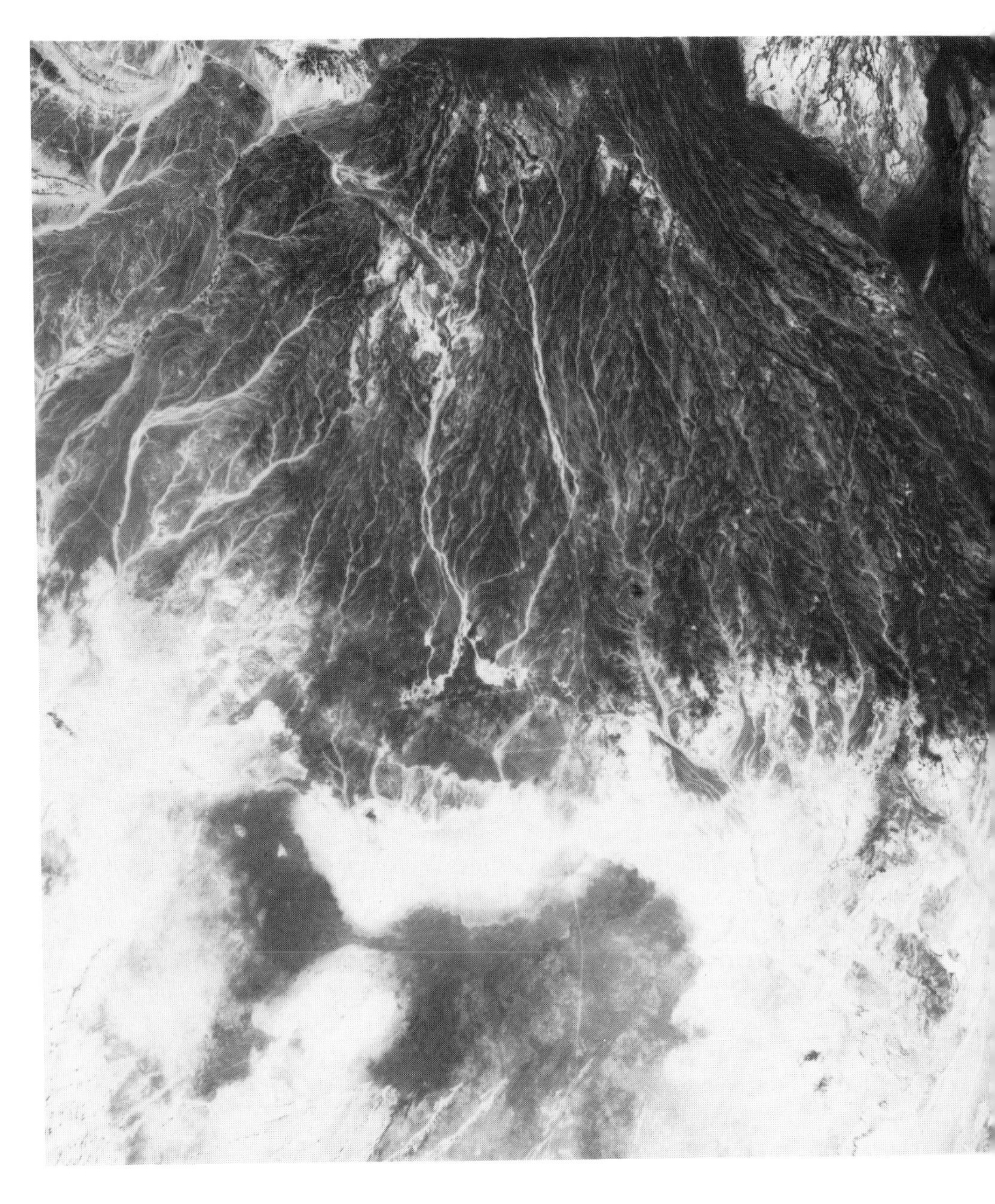

Fig. 4a. Landsat scene of the Oman mountains on the north flank of the interior basin of Oman.

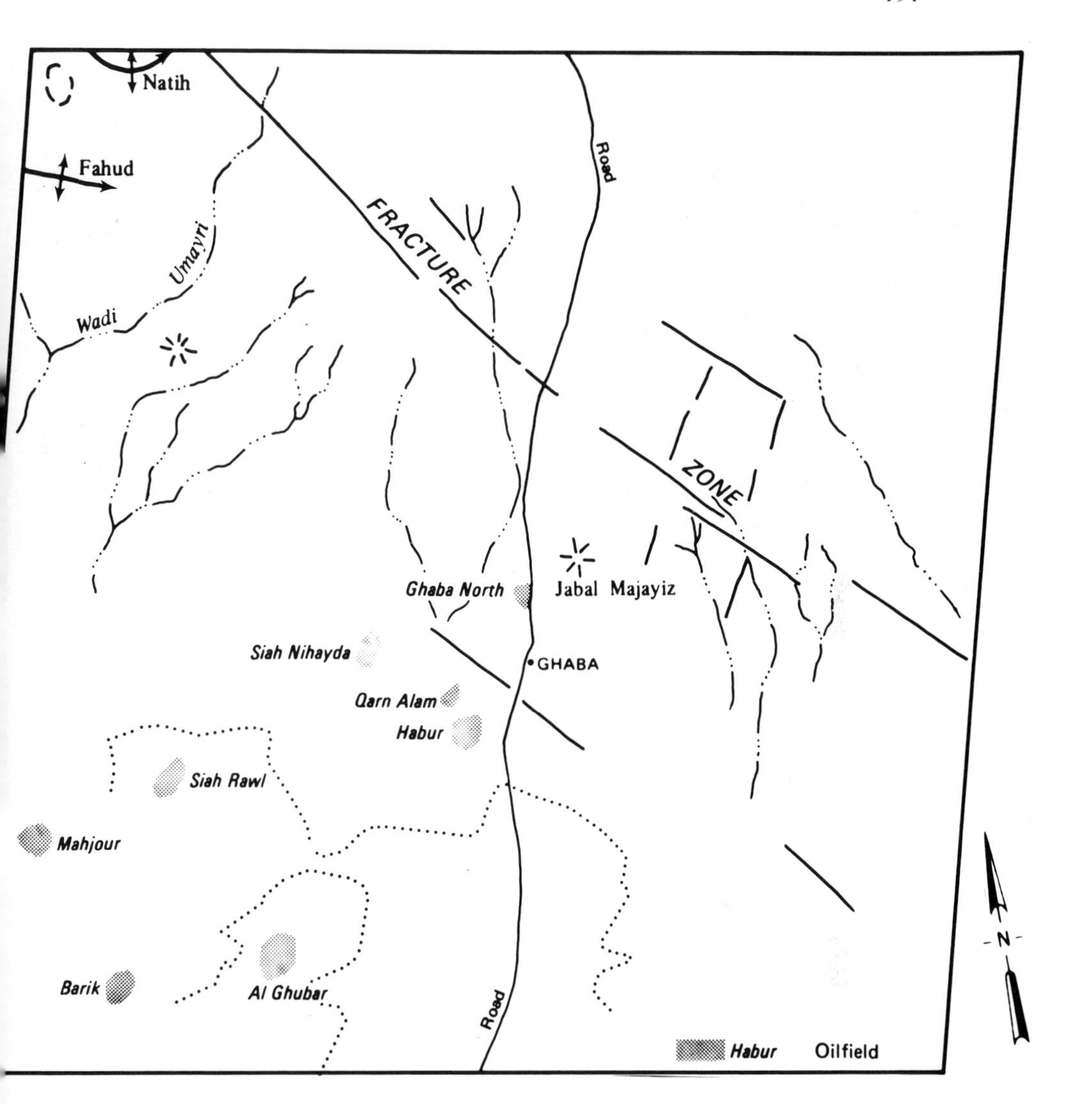

Fig. 4b. Drawing of the same Landsat image showing the fracture zone represents relatively recent faulting based on the fresh escarpments. Oil production is associated with the anticlinal features Natih and Fahud. The several features present in the lower part of the image are all associated with salt doming. This area accounts for most of the oil production of Oman. Jabal Majayiz and the circular feature near Wadi Umayri are probably salt diapirs which have breached the surface. Wadi Umayri, in this 1972 image shows extensive vegetation growth along much of its course, indicating that substantial quantities of shallow groundwater are being lost to the desert.

can be seen in the alluvial plains south of the Oman Mountains and anticlinal structures south of Nizwa show remarkable similarity to the salt cored anticlines in southern Iran (Fig. 2). These structures, Jabal Salakh, Jabal Madmar and Jabal Madar, appear likely to be still forming. Jabal Madar is reported [7] to have salt exposed at the surface. About 80 km south of Ibri, similarly appearing anticlinal structures are associated with some of the major oil production in Oman.

The other area of major production in northern Oman lies near the eastern end of the broad interior depression or structural basin south of the Oman Mountains, mentioned in the Groundwater Section. At least eight fields spread over an area of about 5,000 km^2 near and to the west of the village of Ghaba are associated with diapir structures. Several of these structures are visible on Landsat imagery, as are other probable diapirs in this general area.

The Recent fault, mentioned above, lies about 50 km northeast of the nearest field (see Fig. 4). From the imagery this faulting, which is interpreted as up-to-the-basin, has displaced drainage channels and fault escarpments which appear to still exist in some areas. Although this fault has little or no influence on the oil accumulation in the 'Ghaba' fields, the Landsat imagery shows several subtly expressed lineaments that are parallel to this fault and that lie at the north edge of the present production. Several northeast trending lineaments are visible in this immediate area which, if they represent faulting or jointing, could form the loci for diapiric movement of the underlying salt.

Several oil fields have been found on the southeast side of the structural basin. Much of the oil here is extremely viscous [8], indicating the loss of the lighter components. The trend of this group of fields is parallel to the coast and they extend over a distance of nearly 200 km. Considerable faulting, also parallel to the coast, can be mapped on Landsat imagery and in the Salalah area fracturing appears more intense closer to the ocean. It can be conjectured that considerable strike-slip movement has occurred along these faults during the Miocene time in association with the movement of the Arabian plate to the northeast. This probability is supported by the presence of Precambrian metamorphic rocks on the southeast side of Jabal Saman (east of Salalah) at the foot of a 600 m high fault escarpment of Tertiary limestone.

Much of the area between Salalah and Masirah Island is relatively free of sand and alluvial cover, allowing greater confidence in the interpretation of the imagery. Fault patterns are complex near the coast in southern Oman extending from South Yemen to near Masirah Island. Fault trends appear to intersect at angles of 20–30°, and some sets appear to closely parallel the coastline, whatever the direction. Several of the oil fields in this area appear within circular anomalies which are topographically positive as they deflect drainage around them. A number of such anomalies can be mapped over a large area on the imagery north and northeast of Salalah. Several of these anomalies lie on major linear features suggesting that faulting may have played a major part in the development of the traps.

Conclusion

The foregoing discussion of the petroleum producing areas of the southeastern Arabian Peninsula as observed on Landsat imagery is, by necessity, generalized. Hopefully, it does convey the fact that through the study and interpretation of Landsat imagery much new information can be derived concerning the regional structure of an area that can be of great value in exploration planning.

Glossary

Erg – term used in North Africa referring to a vast region in the Sahara Desert deeply covered with shifting sand and occupied by complex sand dunes; an extensive tract of sandy desert; a sand sea.
Falaj (aflaj, pl.) – an underground water system in which wells are dug to the water table and horizontal tunnels, or qanats, channelize the water to the surface down slope where it is available for use.
Ground truth – information obtained on the surface or subsurface features used to aid in interpretation of remotely sensed data.
Nebkha – small dunes formed on the lee side of shrubs or tussocks of grass. These dunes can develop to a few meters in height.
Phraetophyte – a deep rooted plant that obtains its water from the water table or the layer of soil just above it.
Playa – a term used for a dried up, vegetation-free, flat-floored area composed of thin, evenly stratified sheets of fine clay, silt, or sand, and representing the bottom (lowermost or central) part of a shallow, completely closed or undrained, desert lake basin in which water accumulates (as after a rain) and is quickly evaporated, usually leaving deposits of soluble salts. It may be hard or soft, and smooth or rough.
Sabkha – area of salt flat which is inundated only occasionally.

References

1. ILACO: Water resources development project. Northern Oman, Interim Report. Netherlands: Arnhem, 1975.
2. Mandaville JP: The scientific results of the Oman flora and fauna survey, 1975: Sultanate of Oman: Ministry of Information and Culture, 1977.

3. U.S. Geol. Survey and American Oil Co., Geologic map of the Arabian Peninsula: Map I-270A, 1963.
4. Glennie KW: Geology of the Oman Mountains. In: Ned. Geol. Mijnbouwka Genoot Verh. v. 31, 423 p, 1974.
5. McKee ED: A Study of Global Sand Seas: USGS Prof. Paper 1052, 1979, 429 pp.
6. World Oil, 1981: Iran-Iraq War Tightens Oil Supplies: 192(3):294, 2/15, 1981.
7. Llewellyn P: (Personal Comm.) Manager, Exploration and Production Div., The British Petroleum Company, Ltd., 1980.
8. Oxford University Press, 1980: The Oxford Map of the Sultanate of Oman: Beirut, Lebanon: Geoprojects.

Authors' address:
Earth Satellite Corporation
7222 47th Street
Chery Chase, MD 20815, USA

11. Measuring spectra of arid lands

Anne B. Kahle

Introduction

Arid lands are the most amenable to spectral remote sensing for earth resources owing to the scarcity of vegetation and the consequently favorable exposures of surfaces. For this reason, most research into the use of multispectral measurements for geologic applications has been undertaken in arid or semi-arid regions. A large amount of multispectral data from laboratory, field, aircraft and satellite instruments now exist and significant progress has been made in the interpretation and use of these data.

Laboratory data have been used to establish both a physical and an empirical basis for remote sensing, allowing theory to be related to observable spectral features of pure and mixed substances, including minerals, rocks and soils. Field acquired spectral data bridge the gap between laboratory data and remote sensing aircraft and satellite scanners. This understanding of the spectral information in remotely sensed image data has led to image processing techniques designed to select and display that information required for a particular problem to be solved. These data have application to a wide variety of problems in geologic mapping and mineral exploration.

Landsat 1–3 data enable us to determine uniquely only the presence or absence of 'limonitic' rocks (iron oxides) and the presence or absence of vegetation. The presence of limonite is inferred from measurements of the Fe^{+3} charge transfer band, whose long wavelength is found between 0.4 and 0.6 μm and the Fe^{+2}–Fe^{+3} electronic transition, whose band lies between 0.8 and 1.0 μm in reflectance spectra. The presence of vegetation is inferred from the chlorophyll structure at 0.5 to 0.65 μm. These determinations are most easily accomplished using band ratioing techniques. While unique identification of other rock materials is not possible, considerable success can be achieved in discriminating among many rock units, and delineating boundaries between them. This allows large areas to be 'mapped' with only minimal field checking to identify the mapped units. These units do not necessarily correspond to traditional geologic units, being based on spectral, and hence compositional differences, rather than differences in geo-

El-Baz, F. (ed.), Deserts and arid lands. ISBN: 90-247-2850-9.

logical age, origin, and lithology. To achieve maximum separation of units it is often desirable to apply one of several different image processing techniques, such as a principal component transformation or construction of color ratio composites.

By extending the wavelength range of scanners further into the infrared than the range measured by the present Landsat satellite scanners (<1.1 μm), it is possible to identify additional surface materials. Using the Thematic Mapper Simulator aircraft scanners, with additional bands centered near 1.6 and 2.2 μm, it is possible to recognize the presence of hydrous minerals from the Al-O-H vibrational overtones of these minerals, located at wavelengths greater than 2.0 μm. Recognition of such hydrous phases sometimes allows mapping of hydrothermal alteration which is usually characterized by the presence of kaolinite, sericite, and possibly other clays. Use of very narrow wavelength bands in the region between 2.0 and 2.5 μm allows specific identification of some of these minerals.

Multispectral emission data in the middle infrared (8–14 μm) show great promise for determining the relative free silica content of rocks based upon the presence of Si-0 vibrational absorption. A combination of reflectance and emittance spectral data from all wavelength regions can be expected to improve greatly the separability and identification of geologic materials.

Scanner image data can also be used to map structural features. Image processing techniques such as spatial filtering or edge enhancement may be used to increase feature recognition. These techniques for use of scanner data are covered in detail in other chapters of this volume.

In this chapter we concentrate on the principles of multispectral remote sensing for geologic applications, and discuss briefly some of the related instruments and techniques for data acquisition and data interpretation. This chapter is intended only as a brief introduction to the subject. The interested reader is encouraged to turn to the numerous excellent papers which discuss the various aspects of the subject in much greater depth.

Theory and laboratory data

Spectral remote sensing is based on the determination of the interaction of electromagnetic radiation with the atoms and molecules making up the surface of the object being sensed. In terms of quantum mechanics, atoms and molecules can exist only in certain allowed energy states. Their energy can change between these allowed energy levels by the absorption or emission of energy in the form of electromagnetic radiation, just equal in amount to the difference in energy between the allowed energy levels. These energy levels are dependant upon the electronic, rotational, vibrational and translational energy of the molecular species, which, except for the translational energy, have only certain allowed values. Because materials have characteristic energy levels, measurement of the absorption or emission of electromagnetic radiation as a function of wavelength, upon interactions with the material, sometimes allows one to infer the type of material present. However, the energy levels of only the simplest of atoms and molecules in a gaseous state can be described exactly by quantum mechanics; the rest are too complicated and can only be approximated. When the atoms or molecules are in solid form, the energy levels are modified by effects of the crystal lattice or ligand field, so description of the energy levels in solids is even more approximate. Nevertheless, spectroscopists are usually able to assign experimentally-observed spectral features in materials to certain transitions between energy levels.

The spectrum of a rock is a composite of the spectra of its constituent minerals, and the spectrum of a soil is a composite of mineral spectra plus the spectra of organic materials and water. The spectral features of minerals, rocks and soils are produced by either electronic or vibrational processes; rotational features are found only in gases where the molecules are free to rotate. Electronic transitions require more energy than vibrational processes and most of these spectral features are located in the ultraviolet and visible range with only a few features, notably due to iron, extending into the infrared [1]. Electronic transitions produce spectral absorption features which are usually very broad. Fundamental vibrational processes give rise to much narrower spectral features in the middle and far infrared (MIR, FIR). None occur at wavelengths shorter than 2.5 μm [2]. The range between 1.0 and 2.5 μm, referred to as the near or short wavelength infrared (SWIR), is the 'overtone' or 'combination tone' region. In this region there are abundant diagnostic spectral features due to the excitation of overtone and combination tone vibrations. Because, in remote sensing, the visible and SWIR are usually observed in the reflection mode and the MIR in the emission mode, we will discuss the two regions separately.

When electromagnetic radiation impinges on a material, energy will be selectively absorbed at those wavelengths corresponding to the energy required to raise the energy state of the constituent particles from one energy level to another. If a continuous spectrum of electromagnetic radiation is reflected from a surface, a depletion of radiation will be observed at these wavelengths. The position and relative strength of these absorption features are the key to material identification.

The most widely used spectral remote sensing technique uses the sun as a source of radiation. The sunlight illuminates the surface, interacts with it, and is reflected into the detector, which measures the intensity as a function of wavelength, so we refer to this as reflectance spectroscopy. In reality the radiation arriving at the surface is scattered by the material in all directions; some penetrates the material where it is preferentially absorbed at wavelengths characteristic of the material, and then is scattered in the direction of the detector. It is this absorption which produces the spectral signature of the material.

The visible and near infrared spectra of rocks and minerals have been discussed by numerous authors. One of the largest collections of such spectra is given in a series of papers by Hunt and Salisbury [1, 3, 4, 5] and by Hunt et al. [6–12] which contain the spectra of over 200 minerals and 150 rock samples. Most of the laboratory spectra illustrated here are originally from that important collection.

The most common components of minerals and rocks, namely silicon, aluminum, and oxygen do not possess energy levels such that transitions between them can yield spectral features in the visible and near infrared range. Consequently, no direct information concerning the bulk composition of geological materials is available in this range. However, considerable indirect information is available because the crystal structure imposes its effect upon the energy levels, and therefore upon the spectra of specific ions present in the structure [13–15].

The electronic energy levels of an isolated ion are usually split and displaced when located in a solid. This is particularly true for the transition metal ions, iron, copper, nickel, cobalt, manganese, chromium, vanadium, titanium, and scandium. These new energy values are primarily determined by the valence state of the ion (for example, Fe^{2+} or Fe^{3+}), and by its coordination number and site symmetry. The spectrum is also affected by the type of ligand formed (for example, metal-oxygen), the extent of distortion of the metal ion site from perfect, and the value of the metal-ligand interatomic distances. Although all the electronic features in the spectra are derived from the transition metal ion, the indirect information provided is about the type of crystal field in which it is located, and it is the type of indirect evidence that provides information concerning the bulk structure of minerals and rocks from crystal field effects [15].

In naturally occurring geological materials most spectral information in the visible and near infrared reflectance spectrum is dominated by the very common presence of iron. Examples of the spectra of some iron-bearing minerals are shown in Figure 1 (after [16]). The rapid fall off in reflected intensity toward the ultraviolet is due to the very strong intervalence charge transfer transition between the ferric and oxygen ions. The features marked with arrows are crystal field transitions.

In the short wavelength infrared, features are observed due to the overtones and combinations involving materials that have very high fundamental frequencies. The most common of these near infrared bands involve the OH stretching mode, and water. In the laboratory spectra of minerals and rocks, whenever water is present two bands appear at 1.4 μm and at 1.9 μm. Other bands will also appear depending upon the form of the water in the mineral. There is only one OH fundamental stretching, which occurs near 2.77 μm. This fundamental OH stretching mode may form combination tones with other

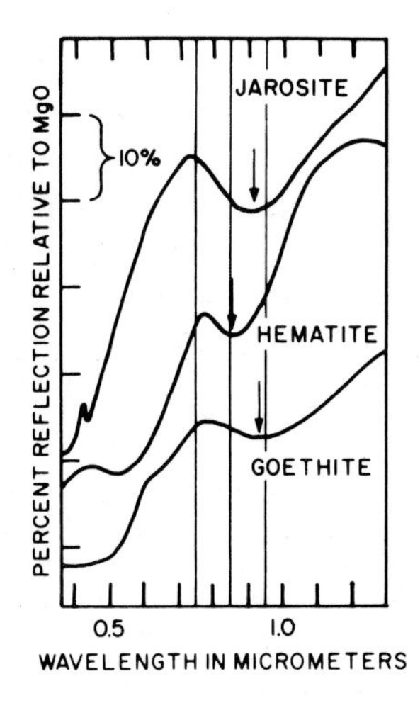

Fig. 1. Visible and near infrared spectra of some iron bearing minerals [16].

fundamentals, including lattice and vibrational modes. In particular, such combinations with fundamental Al-OH or Mg-OH bending modes produce features near 2.2 or 2.3 μm, respectively. Figure 2 [15] displays typical spectra of minerals that contain hydroxyl groups (muscovite, kaolinite) or water (montmorillonite, gypsum, and quartz). The vertical lines mark the centers of the various spectral features.

In addition to the water and OH features, the carbonate minerals display somewhat similar features between 1.6 and 2.5 μm, which are due to combinations and overtones of the four fundamental internal vibrations of the planar CO_3^{-2} ion. These fundamental vibrations produce five features in the near infrared near 1.9 μm; 2.0 μm; 2.16 μm; 2.35 μm; and 2.55 μm [15]. Figure 3, after Hunt and Salisbury [3], shows the spectra of some carbonates. In addition to the carbonate bands apparent in all four spectra, the magnesite and rhodochrosite have a broad ferrous ion band centered near 1.1 μm.

In the middle infrared (MIR) (5–40 μm) region of the spectrum, the emission spectrum of the surface material can be measured. Electromagnetic black-body radiation being emitted as a continuous spectrum will be differentially absorbed before exiting the material.

Laboratory measurements of middle infrared spectra of rocks and minerals show many diagnostic features. The region between 8 and 14 μm holds the most promise for remote sensing because this is an excellent spectral window in the Earth's atmosphere and is also the region of maximum thermal emission at terrestrial surface temperatures [17]. Within this spectral range, the most prominent spectral features are due to silicon-oxygen stretching vibrations. These features change wavelength and intensity with varying composition and struc-

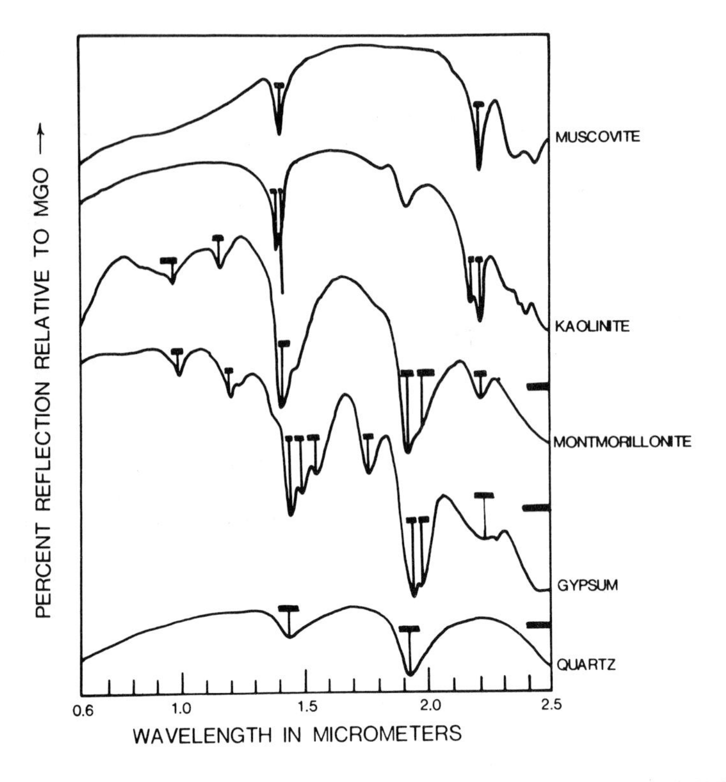

Fig. 2. Visible and near infrared spectra of some hydroxyl and water bearing minerals [15].

ture [18–20, 2, 21]. Lyon and Patterson [22] indicated that these features (known as 'Reststrahlen' bands in reflectance spectra of polished samples) shift to shorter wavelengths as silica content increases.

The most complete set of laboratory spectra of rocks available for the MIR are the transmission and reflection spectra of Hunt and Salisbury [20, 2, 19]. It is important to note that the transmission spectra were obtained using two different sample preparation techniques, the conventional method of the powdered specimens being compressed in KBr pellets, and by deposition of thin layers of fine particles onto a mirror. The principal Christiansen peaks [23] appear in the spectra of the deposited particles (with the particles on the mirror in air), but not in the spectra obtained from the samples in the KBr pellets. This is because the Christiansen peaks, which are located at or near the wavelength at which the refractive index of the material matches that of the medium in which it exists, are minimized in the KBr pellets where the refractive index is closer to the rock samples than is the refractive index of air. However, apart from this difference, most of the absorption features are similar. The reflection spectra measured using polished rock surfaces display more significant differences. Emission

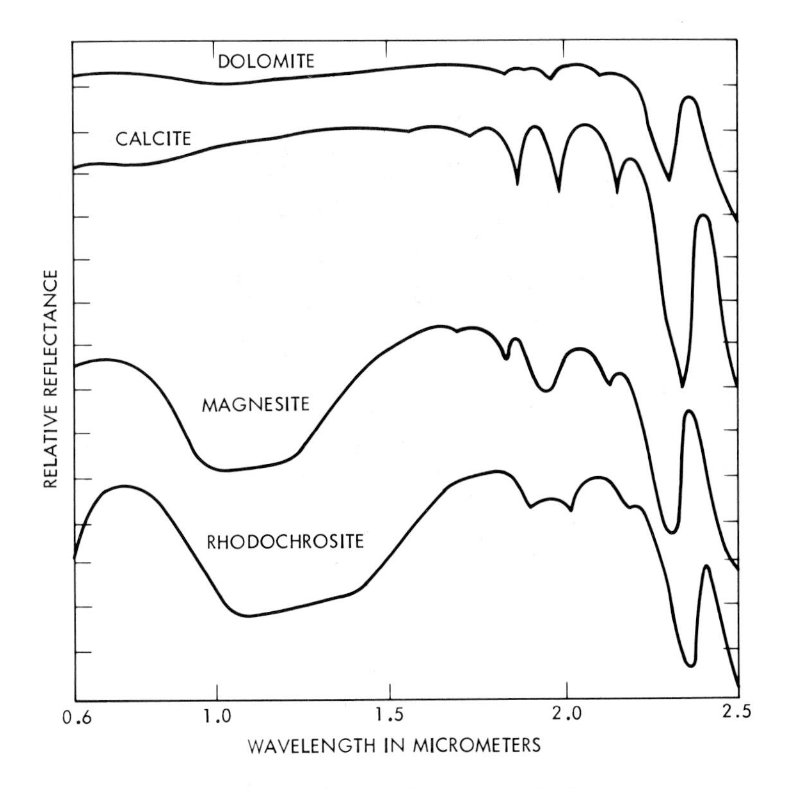

Fig. 3. Visible and near infrared spectra of some carbonate minerals [3].

spectra, which would be the most applicable to remote sensing of emitted radiance are more difficult to measure and interpret. Most of the differences between the spectra obtained in the various modes have been shown both theoretically and experimentally [23, 24, 25] to be dependant upon such variables as particle size, surface roughness, packing density, and near-surface temperature gradients. These parameters affect both the intensity and location of spectral features. Nevertheless, we use the laboratory spectra as a key to interpreting the spectra of natural surfaces. In many cases there are a sufficient number of identifiable spectral features for mineral recognition. In Figure 4 we show a few of the laboratory spectra of Hunt and Salisbury [19, 20, 2] illustrating the features likely to be detected by remote sensing. In the silicate rocks (quartzite, quartz monzonite, monzonite, latite, olivine basalt), the broad, deep decrease in transmission between 8 and 11 μm has been identified by Hunt and Salisbury as being due to Si-O stretching vibrations. The depth and position of the band have been shown to be related to the structure of the constituent minerals and are especially sensitive to the quartz content of the rocks. In the carbonate rocks (limestone, dolomite), the most prominent feature is the C-O

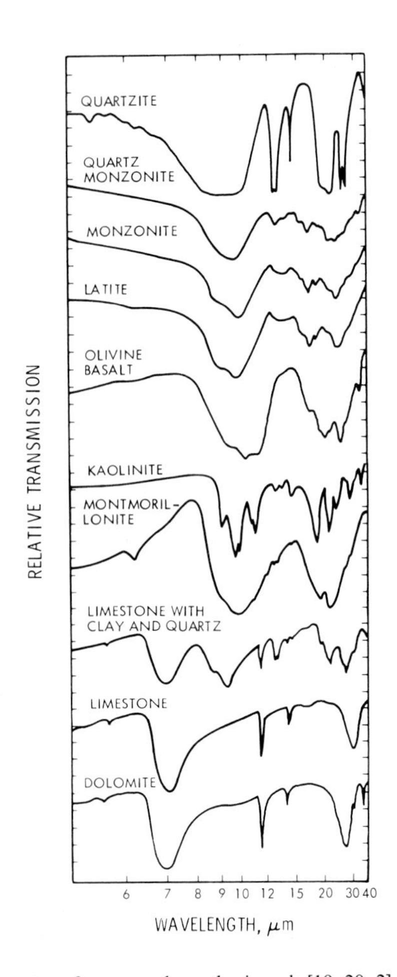

Fig. 4. Middle infrared spectra of some rocks and minerals [19, 20, 2].

asymmetric stretch which, near 7 μm, lies outside the 8 to 14 μm atmospheric window. Throughout the 8–14 μm region, the carbonate spectra are almost flat or show a gradual rise except for the narrow, out-of-plane bending feature near 11.5 μm. Spectral features between 8 and 14 μm in the clays (kaolinite and montmorillonite) are ascribed by Hunt and Salisbury to various Si-O-Si and Si-O stretching vibrations and an Al-O-H bending mode. The montmorillonite features are less distinct than those in the kaolinite, because the numerous exchangeable cations and water molecules in the montmorillonite structure allow many different vibrations.

Field data

Acquisition of spectra of undisturbed surfaces in the field is an important step between the understanding of laboratory spectra of carefully prepared small samples and the interpretation of airborne or satellite spectrometer or multispectral scanner data of large areas of the natural surface materials. Several field portable instruments are now commercially available, with a variety of different wavelength ranges, spectral resolutions, fields of view, sensitivity, and data recording systems. The data from three experimental systems developed at the Jet Propulsion Laboratory will be described here, the Portable Field Reflectance Spectrometer (PFRS), the Hand-Held Ratioing Radiometer (HHRR), and the Portable Field Emission Spectrometer (PFES).

The PFRS [26], see Figure 5, is a self-contained portable field instrument which measures surface reflectance in the wavelength region from 0.442 to 2.53 μm with moderate resolution ($\Delta\lambda/\lambda = 0.04$ from 0.44 to 0.7 μm and 0.015 from 0.7 to 2.53 μm). The instrument utilizes a circular variable filter and a cooled lead-sulfide detector. With the tripod-mounted optical head approximately 1.3 m above the surface, the field of view is rectangular and covers 200 cm^2. The reflectance spectrum of a natural surface illuminated by the sun is taken, followed immediately by the taking of a spectrum of a standard (fiberfrax, a ceramic wool) placed over the surface in the same orientation. Each scan takes thirty seconds and the data are recorded on a digital cassette in a backpack containing the recorder, instrument electronics and power supply. By taking a point-by-point ratio of the spectra of the surface and the standard, a bidirectional reflectance spectrum, independent to the first order of source and atmospheric conditions and surface attitude, is obtained.

Some typical spectra taken using the PFRS are shown in Figure 6. These illustrate most of the features discussed in the previous section. The gaps at 1.4 and 1.9 are imposed because strong atmospheric water absorption bands reduce the incident solar radiation below a useable level. Curve 1 is the spectrum of unaltered tuff fragments and soil and can be seen to be spectrally quite flat. Curves 2 and 3 are argillized andesite and silicified dacite. Both spectra have iron absorption features around 0.9 μm and strong OH absorption near 2.2 μm. The spectrum of an opaline tuff, curve 4, shows a much weaker iron absorption with a still fairly strong OH feature. A typical carbonate – a tan marble is shown in curve 5, with a strong carbonate band near 2.35. Finally, curve 6, the spectrum of ponderosa pine, illustrates the chlorophyll absorption features in the visible and near IR, with the sharp rise in reflectance around 0.7 μm typical of most vegetation. Thousands of field spectra of surface materials have now been obtained [27]. Use of these data has demonstrated that the spectral features noted in the laboratory are indeed present in naturally occurring materials in situ in sufficient strength to be used for remote sensing for geologic applications [28–30].

Fig. 5. The Portable Field Reflectance Spectrometer.

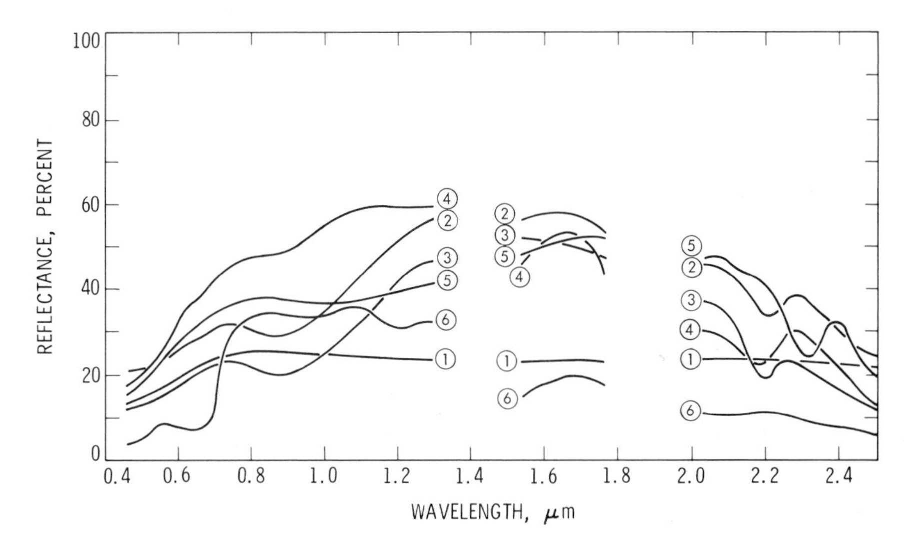

Fig. 6. Field-acquired reflectance spectra of (1) unaltered tuff fragments and soil; (2) argillized andesite fragments; (3) silicified dacite; (4) opaline tuff; (5) tan marble; and (6) ponderosa pine [35].

The Hand-Held Ratioing Radiometer (HHRR) has been developed for use in identifying rocks and minerals in the field (see Fig. 7). This small field instrument is an outgrowth of analysis of laboratory spectra and field spectra from the PFRS. The HHRR is a self-contained, dual-beam, ratioing radiometer with digital readout. It contains two optical trains, each containing two repeater lenses and a cooled lead-sulfide detector. One of the trains is adjustable so that measurements can be made from distances of 1 meter to infinity. The optical trains are intersected by a set of two, coaxially-mounted filter wheels each containing five interference filters. Filters with band passes as narrow as 0.01 μm can be used in the region 0.4 to 2.4 μm. The instrument measures the surface radiance in the two channels and simultaneously displays the ratio.

There is a need for geologists to identify and rapidly map the composition of surface materials. In particular, clay minerals and other OH^- bearing silicates can be difficult to identify in the field. The normal procedure is to return the samples to the laboratory for X-ray analysis. As shown in the previous section, clays and carbonates have unique spectral properties in the 2–2.5 μm region. The HHRR provides a means to identify a number of mineral and rock types. By using properly chosen spectral filters, and ratioing the signals to remove the effect of topography (i.e. directional effects) on the brightness measured, a number of materials can be identified uniquely. Table 1 lists one choice of filters used in HHRR (and also in the Shuttle Multispectral Infrared Radiometer (SMIRR) discussed in the last section of this chapter). Ratio values are shown in Table 2 for several minerals, using these filters. It can be seen, for instance that

Fig. 7. The Hand-Held Ratioing Radiometer.

Table 1. Spectral bands for HHRR and SMIRR.

Channel	Center (μm)	Half-power Bandwidth (μm)
1	0.5±.02	0.1
2	0.6±.02	0.1
3	1.05±.02	0.1
4	1.2±.02	0.1
5	1.6±.02	0.1
6	2.1±.02	0.1
7	2.17±.005	0.02
8	2.20±.005	0.02
9	2.22±.005	0.02
10	2.35±0.15	0.06

Table 2. Normalized band-ratios of calculated responses to laboratory reflectance data for HHRR filter bandpasses.

Band-ratio	Alunite	Kaolinite	Mont-morillonite	Calcite
2.10/2.20	1.11	1.35	1.26	1.00
2.10/2.22	0.93	1.25	1.53	1.01
2.17/2.20	0.86	1.07	1.32	1.01
2.17/2.22	0.72	0.98	1.59	1.01
2.17/2.35	0.72	1.02	2.02	1.11

montmorillonite and kaolinite yield very different ratio values for filters centered at 2.10 and 2.17 μm. For a number of materials, HHRR can thus be used as an analytical tool in the field. The instrument can be used on the ground or from a moving vehicle or aircraft.*

The Portable Field Emission Spectrometer (PFES) used the same electronics and digital recording backpack as the PFRS, with a new filter wheel and detector system allowing emission spectra to be measured in the 3–5 μm and 8–14 μm ranges. Three PFES spectra of monzonite, quartz monzonite, and quartz are shown in Figure 8. Here we see the Si-O absorption feature becoming progressively deeper with increased quartz content of the rock. The instrument will continue to be used to test the feasibility of using measurements of emission spectra of naturally occurring materials as a remote sensing technique.

* HHRR is now being manufactured by Barringer Research, Inc., Golden, Colorado, U.S.A.

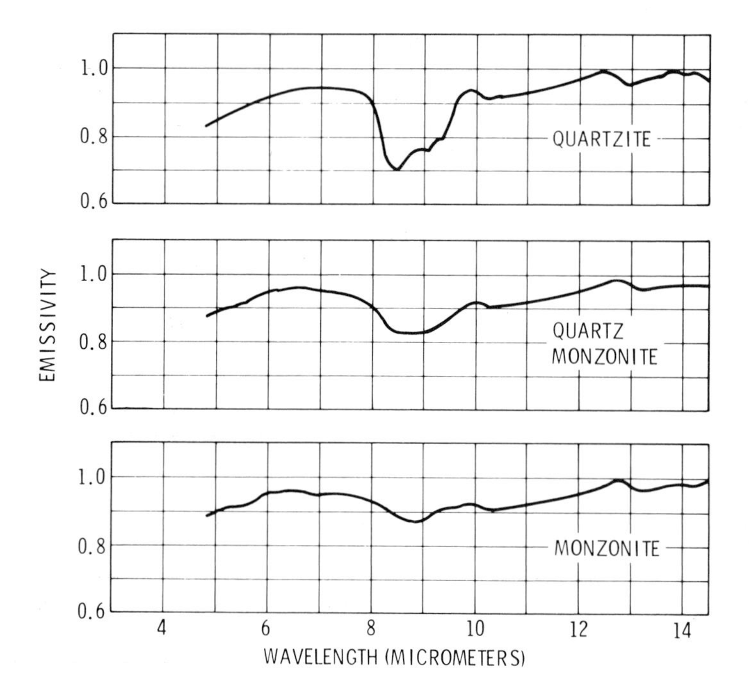

Fig. 8. Field-acquired emission spectra.

Aircraft and satellite spectral measurements

Aircraft and satellite instruments observing the Earth in the visible and infrared wavelengths, exist both as profilers, taking measurements in a series of points lying on a line along the track of the aircraft or spacecraft, or as scanners, building up an image of a scene by scanning back and forth across the tract while allowing the motion of the craft to create the second dimension of the image. Radiometers measure the radiance from one, or possibly more discrete broad band portions of the spectrum, while spectrometers measure data from a few medium width wavelength bands or up to hundreds of narrow wavelength bands or channels.

Aircraft data can be used as an intermediate step between field acquired data and data acquired from orbiting satellite systems. These data can be used both to prove that a remote sensing technique will work prior to the construction and deployment of expensive orbiting systems, and to examine in more detail areas which appear, on satellite data, to be of interest but where better spatial detail is required. Satellite data have the advantages of covering large portions of the Earth, often many times, under nearly uniform viewing conditions.

Data from the Landsat satellite system and from three of NASA's experimental multispectral aircraft scanners will be discussed here, to illustrate some of the geologic information that can be obtained in arid regions from such instruments. Interpretation of the image data is based on an understanding of the spectral characteristics of the surface materials as discussed in the previous sections.

By far the best known and most widely used multispectral satellite data of the Earth are from the Landsat satellites. These are a series of satellites (four to date, with more planned) launched at intervals so as to provide continuous coverage from 1972 onward. The instrument on the first four satellites of interest here is the Multispectral Scanner (MSS) which has four channels in the spectral region 0.5 to 0.6 μm (channel 4), 0.6 to 0.7 μm (channel 5), 0.7 to 0.8 μm (channel 6) and 0.8 to 1.1 μm (channel 7). The data are acquired over a swath width of 185 km, from a sun-synchronous orbit, with repeating coverage of the Earth every 18 days. The instantaneous field of view (or picture element, pixel, size) is approximately 60 m $\times$ 80 m. The data are available in either digital or image format from the EROS Data Center in Sioux Falls, South Dakota. In addition, on Landsat 4, the new Thematic Mapper was also flown with wavelength channels as shown in Table 3.

The use of Landsat data to study landforms of arid regions is described in other chapters. The discussion here is concerned with the geologic information that can be derived from spectral MSS data. Photographic Landsat data products have some use in lithologic discrimination. A black and white image from a single MSS channel will allow some separation of rock units based on their brightness and textural differences. Color composites of three bands will allow better separation, based on spectral (color) differences between units. However, the best success is achieved if the Landsat data are digitally processed to enhance the particular features of interest. Color ratio composites have proven to be very useful in emphasizing spectral differences between units while subduing the unwanted brightness differences related to topographic effects [31, 26, 32]. Because the MSS bands are situated in the visible and near infrared regions of the spectrum between 0.5 to 1.1 μm, these color ratio composites are well suited for identifying the presence of iron oxides, as discussed in the previous sections. The 'limonitic' rocks are often associated with hydrothermal alteration which may be related to economically important mineralization. However such color composites will also indiscriminately indicate areas where iron oxides have no relationship to hydrothermal alteration such as iron-oxide rich volcanic or sedimentary rocks or some areas of weathered alluvium. Additionally, areas of economically important hydrothermal alteration may not happen to contain significant limonitic material so they will fail to be identified by the MSS. Possible solutions to resolving some of these ambiguities can be found by looking at either narrower wavelength band data to help identify particular iron oxides, or by examining data from longer wavelengths in the SWIR as provided

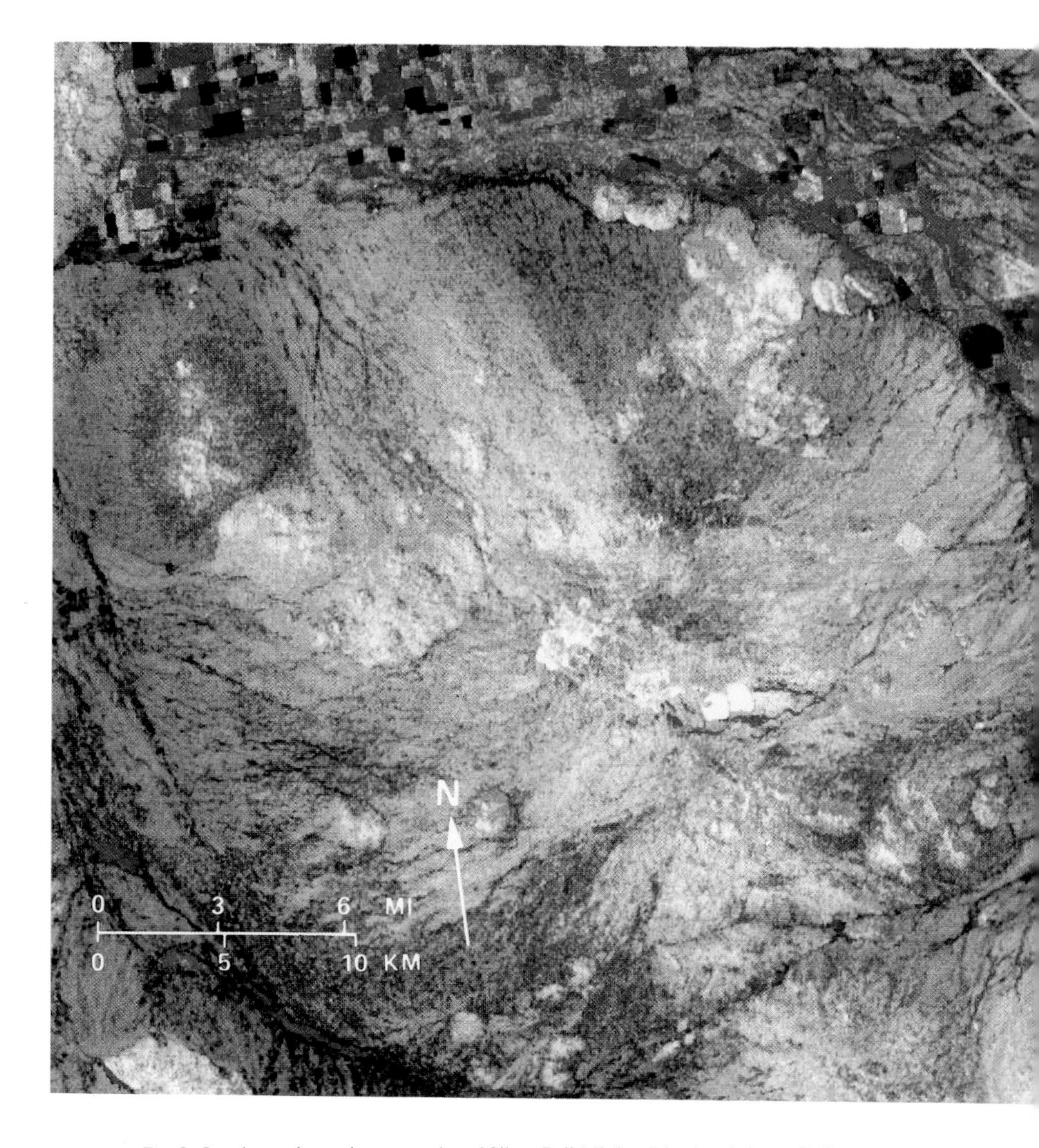

Fig. 9. Landsat color ratio composite of Silver Bell Mining District, Arizona [36].

by the Thematic Mapper, to look for evidence of the OH bands of the layered silicate minerals, which are also often associated with hydrothermal alteration [28, 33–35].

An example of an Landsat MSS color ratio composite is shown in Figure 9 [36]. This is an image of the Silver Bell mining district, near Tucson, Arizona in the Southwestern desert region of the USA. Band ratios 4/5, 5/6 and 6/7 are

Fig. 10. Thematic Mapper Simulator color ratio composite of Silver Bell Mining District, Arizona [36].

displayed in blue, green, and red respectively. The two irregular shaped yellow areas near the center are two open pit porphyry copper mines. The nearby yellow squares are leaching pits. The orange areas north of the leach pits, around the pits, south of the right pit and in the right corner all contain iron-oxide bearing materials. Some of these rocks are hydrothermally altered but some are unaltered sedimentary red beds.

Aircraft flights were made in this area with two of NASA's multispectral scanners. One, the M^2S 11-channel scanner has 10 channels between 0.4 and 1.1 μm and one middle IR channel, and the other scanner, the Thematic Mapper Simulator has the 8 channels shown in Table 3. (The Thematic Mapper Simulator is an aircraft equivalent (simulator) of the Thematic Mapper, the new scanner on Landsat 4.)

Analysis of color ratio composite images from the M^2S 11-channel scanner at other nearby mining districts suggests that hematitic areas could be separated from limonitic/goethitic areas because of the narrowness of the bands on this particular scanner. It is possible that the shift of the Fe^{+3} absorption band from 0.85 to 0.95 due to the change from hematite to limonite/goethite was being detected, but this still needs to be verified in the field [36].

This same scanner, though, showed essentially no new information at Silver Bell other than improved spatial resolution. A great improvement was realized at Silver Bell, however, with data from the Thematic Mapper Simulator. These data are shown in Figure 10 with the ratios 0.83/1.15, 0.66/0.56, and 1.65/2.22 displayed in blue, green and red. These ratios were chosen so that iron-oxide minerals would appear green and hydrous minerals red. Materials with both of these substances present should then appear yellow. The yellow areas do indeed correspond to the field mapped areas of hydrothermal alteration. The sedimentary red beds south of the pits, which were indistinguishable from the altered rocks on the Landsat MSS image, now appear green and hence are clearly distinguishable [36].

As previously mentioned, the other limitation on the ability of Landsat MSS data to identify areas of hydrothermal alteration, i.e., areas with hydrous minerals but no iron oxides, has also been shown to be identifiable if the correct wavelength channels are available. NASA's 24-channel aircraft scanner was flown over Cuprite, Nevada, also in the American Southwestern Desert. This scanner had channels in the visible and near IR very similar in wavelength to those on the Thematic Mapper Simulator, including those centered near 1.6 and 2.2 μm. Here, despite the lack of significant iron oxides, color ratio composites clearly delineated areas which were bleached and highly altered, even separating into various grades of alteration [37].

The use of spectral emittance data in the middle infrared (MIR) for mineral identification has also been demonstrated wi:h NASA's 24-channel scanner data [38, 39] following earlier demonstrations with 2-channel data [40–42]. The 24-

Table 3. Thematic Mapper (TM) and Thematic Mapper Simulator (TMS) channels.

TMS Channel	TM	Wavelength (μm)	TMS Channel	TM	Wavelength (μm)
1	1	0.45–0.52	5	–	1.00–1.30
2	2	0.52–0.60	6	5	1.55–1.75
3	3	0.63–0.69	7		2.08–2.35
4	4	0.76–0.90	8		10.4–12.5

channel scanner had 6 channels between 8 and 13 μm. A principal component color composite image using three bands is shown in Figure 11, from data acquired over Tintic, Utah, a semi-arid geologically complex area of high relief and moderate vegetation. In general, the red colors represent rocks in which quartz is a major constituent, while green indicates non-silicate rocks (carbonates) and vegetation. The more intense the red, the higher the quartz content of a unit, illustrating the dominant effect of the Si-O spectral features in this wavelength region. It is significant, for instance, that quartz monzonite and quartz latite appear pinkish in the image while monzonite and latite appear blue. This image helps demonstrate the promise this spectral region holds for geologic applications.

Ongoing Research and Future Systems

Based on the encouraging results to date of multispectral remote sensing for geologic applications, several new experimental systems are in operation, and others are planned.

As mentioned earlier, the Thematic Mapper with bands similar to those on the current MSS, and with the addition of the important bands centered near 1.6 and 2.2 μm has been flown on Landsat 4. In addition to the new spectral channels, the instrument has improved spatial resolution, 30 m rather than the 80 m of the MSS. First results are promising.

An experimental narrow band profiling instrument, the Shuttle Multispectral Infrared Radiometer (SMIRR), was flown on the second Space Shuttle flight [44]. This instrument had 10 medium and narrow band filters (see Table 1), similar to the field instrument HHRR described earlier, and was designed to test and successfully demonstrated the feasibility of using narrow band spectrometry for mineral identification from space. An Airborne Imaging Spectrometer (AIS) has been developed that is capable of acquiring images in 128 spectral bands simultaneously in the region 1.2–2.4 μm. Future imaging spectrometer systems are being developed for high altitude aircraft (Airborne Visible Infrared Imaging

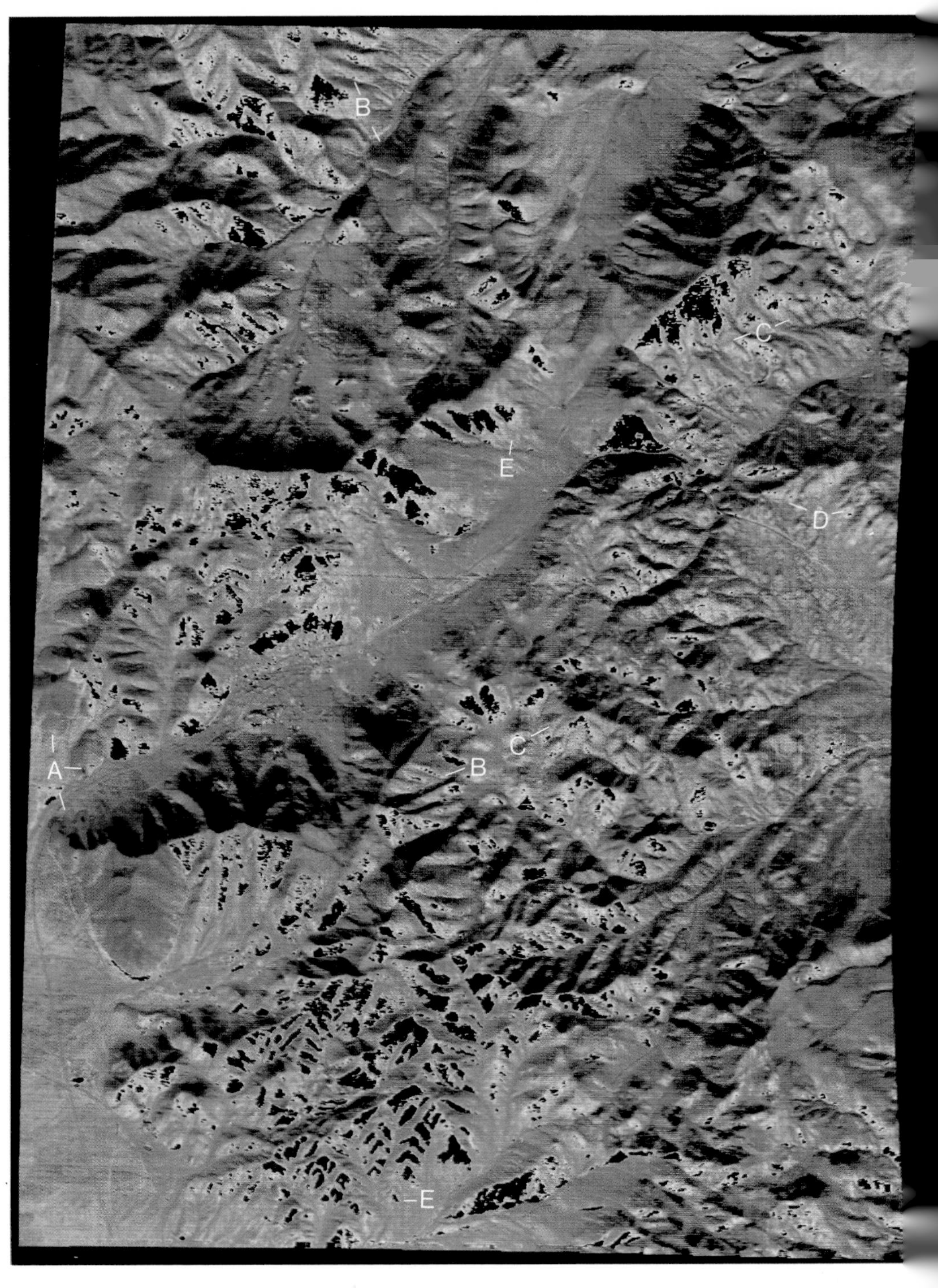
B
C
E
D
A
C
B
E

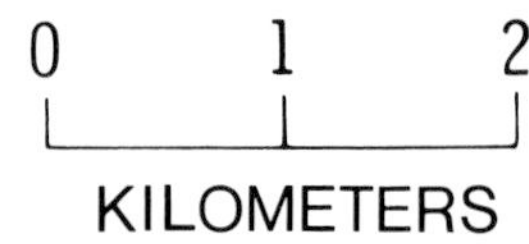
0
1
2
KILOMETERS

←

Fig. 11. Color composite image of the East Tintic Mountains, Utah, based on principal component transformations of middle infrared multispectral data, with (A) quartz rich rocks; (B) interlayered quartz-rich and carbonate rocks; (C) silicified rocks; (D) quartz latite and quartz monzonite; (E) latite and monzonite; and (F) areas that exceeded the thermal response range of the scanner [38].

Spectrometer (AVIRIS)) and shuttle (Shuttle Imaging Spectrometer Experiment (SISEX)). The imaging spectrometer systems have the advantage that for each pixel in the image it is possible to derive a complete reflectance spectrum. SMIRR is a precursor to narrow band imaging systems.

A new aircraft scanner, the Thermal Infrared Multispectral Scanner (TIMS), has been completed with six bands in the 8 to 13 μm range, comparable to the (no longer operational) 24-channel scanner. The first data from this instrument [45] have proven to be as useful geologically as anticipated from the 24-channel data results, and it is considered as a precursor to an orbiting system.

As the number of wavelength channels identified to be important becomes larger, with more and narrower channels, and as increased spatial resolution is demanded, new scanner and data processing technology must be developed.

References

1. Hunt GR, Salisbury JW: Visible and near-infrared spectra of minerals and rocks: I silicate minerals. Modern Geology (1):283–300, 1970.
2. Hunt GR, Salisbury JW: Mid-infrared spectral behavior of metamorphic rocks, Environ. Res. Paper B43-AFCRL-TR-76-0003, 1976, 49 pp.
3. Hunt GR, Salisbury JW: Visible and near-infrared spectra of minerals and rocks: II carbonates. Modern Geology (2):23–30, 1971a.
4. Hunt GR, Salisbury JW: Visible and near-infrared spectra of minerals and rocks: XI sedimentary rocks. Modern Geology (5):211–217, 1971b.
5. Hunt GR, Salisbury JW: Visible and near-infrared spectra of minerals and rocks: XII metamorphic rocks. Modern Geology (5):221–228, 1971c.
6. Hunt GR, Salisbury JW, Lenhoff CJ: Visible and near-infrared spectra of minerals and rocks: III oxides and hydroxides. Modern Geology (2):195–205, 1971a.
7. Hunt GR, Salisbury JW, Lenhoff CJ: Visible and near-infrared spectra of minerals and rocks: IV sulphides and sulphates. Modern Geology (3):1–4, 1971b.
8. Hunt GR, Salisbury JW, Lenhoff CJ: Visible and near-infrared spectra of minerals and rocks: V halides, phosphates, arsenates, vanadates, and borates. Modern Geology (3):121–132, 1972.
9. Hunt GR, Salisbury JW, Lenhoff GJ: Visible and near-infrared spectra of minerals and rocks: VI additional silicates. Modern Geology (4):85–106, 1973a.
10. Hunt GR, Salisbury JW, Lenhoff CJ: Visible and near-infrared spectra of minerals and rocks: VII acidic igneous rocks. Modern Geology (4):217–224, 1973b.
11. Hunt GR, Salisbury JW, Lenhoff CJ: Visible and near-infrared spectra of minerals and rocks: VII intermediate igneous rocks. Modern Geology (4):237–244, 1974a.

12. Hunt GR, Salisbury JW, Lenhoff CJ: Visible and near-infrared spectra of minerals and rocks: IX basic and ultrabasic rocks. Modern Geology (5):15–22, 1974b.
13. Burns RG: Mineralogical applications of crystal field theory, London: Cambridge University Press, 1970.
14. Hunt GR: Spectral signatures of particulate minerals in the visible and near-infrared. Geophysics (42):501–513, 1977.
15. Hunt GR: Electromagnetic Radiation: The communication link in remote sensing. In: Siegal BS, Gillespie AR (eds) Remote Sensing Geology. New York: John Wiley and Sons, 1980.
16. Hunt GR, Ashley RP: Spectra of altered rocks in the visible and near infrared. Econ. Geol. (74):1613–1629, 1979.
17. Vincent RK: The potential role of thermal infrared multispectral scanners in geologic remote sensing. Proc. IEEE (63):137–147, 1965.
18. Lyon RJP: Analysis of rocks by spectral infrared emission (8 to 25 microns). Econ. Geol. (60):715–736, 1965.
19. Hunt GR, Salisbury JW: Mid-infrared spectral behavior of igneous rocks. Environ. Res. Paper 496-AFCRL-TR-74-0625, 1974, 142 pp.
20. Hunt GR, Salisbury JW: Mid-infrared spectral behavior of sedimentary rocks. Environ. Res. Paper 520-AFCRL-TR-75-0256, 1975, 49 pp.
21. Vincent RK, Rowan LC, Gillespie RE, Knapp C: Thermal-infrared spectra and chemical analyses of twenty-six igneous rock samples. Remote Sensing of Environment (4):199–209, 1975.
22. Lyon RJP, Patterson JW: Infrared spectral signatures – A field geological tool. Proc. of the Fourth Symposium on Remote Sensing of Environment, 1966, pp 215–230.
23. Conel JE: Infrared emissivities of silicates: experimental results and a cloudy atmosphere model of spectral emission from condensed particulate mediums. J. Geophys. Res. (74):1614–1634, 1969.
24. Emslie AG, Aronson JR: Spectral reflectance and emittance of particulate materials 1: theory, Appl. Optics (12):2563–2572, 1973.
25. Aronson JR, Emslie AG: Spectral reflectance and emittance of particulate materials 2: application and results. Appl. Optics (12):2573–2584, 1973.
26. Goetz AFH, Billingsley FC, Gillespie AR, Abrams MJ, Squires RL, Shoemaker EM, Jucchitti I, Elston DP: Application of ERTS image processing to regional geologic problems and geologic mapping in northern Arizona, Pasadena, California: Jet Propulsion Laboratory, Technical Report 13–1597, 1975.
27. Kahle AB, Goetz AFH, Paley HN, Alley RE, Abbott EA: A data base of geologic field data. Proc. Fifteenth International Symposium on Remote Sensing, Ann Arbor, Michigan, 1981.
28. Rowan LC, Goetz AFH, Ashley RP: Discrimination of hydrothermally altered and unaltered rocks in visible and near-infrared multispectral images. Geophysics v. 42, n. 3:522–535, April, 1977.
29. Conel JE, Abrams MJ, Goetz AFH: A study of alteration associated with uranium occurrences in sandstone and its detection by remote sensing methods, Pasadena, California: Jet Propulsion Laboratory, Report No. 78–66, 1978.
30. Blom R, Abrams MJ, Adams HG: Spectral reflectance and discrimination of plutonic rocks in the 0.45 – to 2.45 – μm region. J. Geophys. Res. (85):2638–2648, 1980.
31. Rowan LC, Wetlaufer PH, Goetz AFH, Billingsley FC, Stewart JH: Discrimination of rock types and altered areas in Nevada by the use of ERTS images. U.S.G.S. Prof. Paper 883, 1974.
32. Gillespie AR: Digital techniques of image enhancement. In: Siegal BS, Gillespie AR (eds) Remote Sensing in Geology. New York: John Wiley and Sons, 1980.
33. Rowan LC, Abrams MJ: Mapping hydrothermally altered rocks in the East Tintic Mountains using 0.4–2.38 μm multispectral scanner aircraft images, abstr. International Association on

the Genesis of Ore Deposits, Snowbird, Alta, Utah, August 1978 program and abstracts, 1978, pp 156.
34. Rowan LC, Lathram EH: Mineral exploration. In: Siegal BS, Gillespie AR (eds). Remote Sensing in Geology, New York: John Wiley and Sons, 1980.
35. Goetz AFH, Rowan LC: Geologic remote sensing. Science (211):781–791, 1981.
36. Abrams MJ, Brown D, Sadowski R, Lepley L: Applications of remote sensing to porphyry copper exploration with emphasis on the proposed Landsat-D Thematic Mapper, 1981. International Geoscience and Remote Sensing Symposium (IGARSS '81) Digest (1):331–336, Washington, DC: IEEE Geoscience and Remote Sensing Society, June 8–10, 1981.
37. Abrams MJ, Ashley RP, Rowan LC, Goetz AFH, Kahle AB: Mapping of hydrothermal alteration in the Cuprite mining district, Nevada, using aircraft scanner imagery for the 0.46–2.36 μm spectral region. Geology v. 5, n. 12:713–718, December 1977.
38. Kahle AB, Rowan LC: Evaluation of multispectral middle infrared aircraft images for mapping in the East Tintic Mountains, Utah. Geology (8):234–239, 1980.
39. Kahle AB, Madura DP, Soha JM: Middle infrared multispectral aircraft scanner data: analysis for geological applications. Appl. Optics: 2279–2290, 1980.
40. Vincent RK, Thomson FJ: Rock-type discrimination from ratioed infrared scanner images of Pisgah Crater, California, Science (175):986–988, 1972.
41. Vincent RK, Thomson F: Spectral compositional imaging of silicate rocks. J. Geophys. Res. (77):2465–2472, 1972a.
42. Vincent RK, Thomson F, Watson K: Recognition of exposed quartz sand and sandstone by two channel infrared imagery. J. Geophys. Res. (77):2473–2477, 1972b.
43. Goetz AFH, Rowan LC: Narrowband IR radiometry for mineral identification: Shuttle multispectral infrared radiometer (SMIRR) aircraft test results, 1981 International Geoscience and Remote Sensing Symposium (IGARSS '81) Digest (1):345–347. Washington, DC: IEEE Geoscience and Remote Sensing Society, June 8–10, 1981b.
44. Goetz AFH, Rowan LC, Kingston MJ: Mineral identification from orbit: Initial results from the Shuttle Multispectral Infrared Radiometer. Science (218):1020–1024, 1982.
45. Kahle AB, Goetz AFH: Mineralogic information from a new airborne thermal infrared multispectral scanner. Science (222) October 7, 1983.

Author's address:
Jet Propulsion Laboratory
Pasadena, CA 91109, USA

Index